IMPLEMENTATION OF
A MANAGEMENT SYSTEM FOR
OPERATING ORGANIZATIONS OF
RESEARCH REACTORS

The following States are Members of the International Atomic Energy Agency:

AFGHANISTAN
ALBANIA
ALGERIA
ANGOLA
ARGENTINA
ARMENIA
AUSTRALIA
AUSTRIA
AZERBAIJAN
BAHRAIN
BANGLADESH
BELARUS
BELGIUM
BELIZE
BENIN
BOLIVIA
BOSNIA AND HERZEGOVINA
BOTSWANA
BRAZIL
BULGARIA
BURKINA FASO
BURUNDI
CAMBODIA
CAMEROON
CANADA
CENTRAL AFRICAN
  REPUBLIC
CHAD
CHILE
CHINA
COLOMBIA
CONGO
COSTA RICA
CÔTE D'IVOIRE
CROATIA
CUBA
CYPRUS
CZECH REPUBLIC
DEMOCRATIC REPUBLIC
  OF THE CONGO
DENMARK
DOMINICA
DOMINICAN REPUBLIC
ECUADOR
EGYPT
EL SALVADOR
ERITREA
ESTONIA
ETHIOPIA
FIJI
FINLAND
FRANCE
GABON
GEORGIA
GERMANY
GHANA
GREECE

GUATEMALA
HAITI
HOLY SEE
HONDURAS
HUNGARY
ICELAND
INDIA
INDONESIA
IRAN, ISLAMIC REPUBLIC OF
IRAQ
IRELAND
ISRAEL
ITALY
JAMAICA
JAPAN
JORDAN
KAZAKHSTAN
KENYA
KOREA, REPUBLIC OF
KUWAIT
KYRGYZSTAN
LAO PEOPLE'S DEMOCRATIC
  REPUBLIC
LATVIA
LEBANON
LESOTHO
LIBERIA
LIBYA
LIECHTENSTEIN
LITHUANIA
LUXEMBOURG
MADAGASCAR
MALAWI
MALAYSIA
MALI
MALTA
MARSHALL ISLANDS
MAURITANIA
MAURITIUS
MEXICO
MONACO
MONGOLIA
MONTENEGRO
MOROCCO
MOZAMBIQUE
MYANMAR
NAMIBIA
NEPAL
NETHERLANDS
NEW ZEALAND
NICARAGUA
NIGER
NIGERIA
NORWAY
OMAN
PAKISTAN
PALAU

PANAMA
PAPUA NEW GUINEA
PARAGUAY
PERU
PHILIPPINES
POLAND
PORTUGAL
QATAR
REPUBLIC OF MOLDOVA
ROMANIA
RUSSIAN FEDERATION
RWANDA
SAUDI ARABIA
SENEGAL
SERBIA
SEYCHELLES
SIERRA LEONE
SINGAPORE
SLOVAKIA
SLOVENIA
SOUTH AFRICA
SPAIN
SRI LANKA
SUDAN
SWAZILAND
SWEDEN
SWITZERLAND
SYRIAN ARAB REPUBLIC
TAJIKISTAN
THAILAND
THE FORMER YUGOSLAV
  REPUBLIC OF MACEDONIA
TOGO
TRINIDAD AND TOBAGO
TUNISIA
TURKEY
UGANDA
UKRAINE
UNITED ARAB EMIRATES
UNITED KINGDOM OF
  GREAT BRITAIN AND
  NORTHERN IRELAND
UNITED REPUBLIC
  OF TANZANIA
UNITED STATES OF AMERICA
URUGUAY
UZBEKISTAN
VENEZUELA
VIETNAM
YEMEN
ZAMBIA
ZIMBABWE

The Agency's Statute was approved on 23 October 1956 by the Conference on the Statute of the IAEA held at United Nations Headquarters, New York; it entered into force on 29 July 1957. The Headquarters of the Agency are situated in Vienna. Its principal objective is "to accelerate and enlarge the contribution of atomic energy to peace, health and prosperity throughout the world".

SAFETY REPORTS SERIES No. 75

IMPLEMENTATION OF A MANAGEMENT SYSTEM FOR OPERATING ORGANIZATIONS OF RESEARCH REACTORS

INTERNATIONAL ATOMIC ENERGY AGENCY
VIENNA, 2013

COPYRIGHT NOTICE

© IAEA, 2013

Printed by the IAEA in Austria
June 2013
STI/PUB/1584

IAEA Library Cataloguing in Publication Data

Implementation of a management system for operating organizations of research
    reactors — Vienna : International Atomic Energy Agency, 2013.
        p. ; 24 cm. — (Safety reports series, ISSN 1020–6450 ; no. 75)
    STI/PUB/1584
    ISBN 978–92–0–136010–6
    Includes bibliographical references.

    1. Nuclear reactors — Research. 2. Nuclear facilities — Management. 3. Nuclear
    facilities — Safety measures. I. International Atomic Energy Agency. II. Series.

IAEAL   13–00817

FOREWORD

Over the years, IAEA requirements to establish and implement a quality assurance programme for research reactors and their associated experiments have been developed, to ensure safe and reliable operation. These requirements range from quality assurance requirements established in 1996 to requirements to establish and implement a process based management system conforming with the requirements set out in IAEA Safety Standards Series No. GS-R-3 on The Management System for Facilities and Activities in 2006. IAEA Safety Standards Series No. NS-R-4 on the Safety of Research Reactors, issued in 2004, states that "the establishment, management, performance and evaluation of a quality assurance programme for a research reactor and its associated experiments are important for ensuring safety."

The IAEA safety standards on quality management of 1996, which were superseded by IAEA Safety Standards Series No. NS-R-4, did not require a process based management system in which all elements of safety, health, environment, security and economics were integrated. As a consequence, the majority of research reactor operating organizations developed different quality assurance programmes for operation, utilization, maintenance, radiation protection and emergency preparedness, resulting in a patchwork of programmes and procedures by which the responsibilities to ensure safe and reliable operation and the safety of workers and the public were distributed among various quality assurance programmes and departments.

With the introduction of IAEA Safety Standards Series No. GS-R-3, it was required that "a management system ... be established, implemented, assessed and continually improved", and that "health, environmental, security, quality and economic requirements are not considered separately from safety requirements, to help preclude their possible negative impact on safety."

However, the need for practical examples of the implementation of process based management systems for operating organizations of research reactors was still apparent, as was also shown by requests from IAEA Member States.

The global research reactor community includes a number of highly diverse facility designs, organizational structures and technical missions. Research reactor power levels range from over 100 MW down to zero power in both critical and subcritical facilities. Some research reactor operating organizations are large and comprise teams of operators, maintenance technicians, safety and radiation control officers, managers, and licensing and other support staff. Other reactors are successfully operated by a relatively small team of around ten or even fewer permanent staff. For this reason, among others, a case study of a smaller research reactor has been included in Annex I of this publication.

This report presents a set of processes which are applicable to many research reactor operating organizations, but the processes will ideally be tailored to the specific operating organization and graded based on the structure of the operating organization and on potential hazards relevant to the specific nuclear facility involved.

The IAEA wishes to thank the senior management of those nuclear organizations that provided relevant information and source materials. Special thanks go to the Nuclear Research and consultancy Group (NRG) based in Petten, the Netherlands, which supplied the basis for the procedures in this publication, and to the University of Pavia, which contributed the case study presented in Annex I. The IAEA would also like to thank the experts from Argentina, Belgium, Canada, Finland, Italy, the Netherlands, the Philippines, Sweden, the United Kingdom and the United States of America who contributed to the preparation of this publication and supplied practical examples. The IAEA officers responsible for this publication were P. Vincze and J. P. Boogaard of the Division of Nuclear Power.

CONTENTS

1. INTRODUCTION

1.1. BACKGROUND

IAEA Safety Standards Series No. NS-R-4 on the Safety of Research Reactors [1], states that "the establishment, management, performance and evaluation of a quality assurance programme for a research reactor and its associated experiments are important for ensuring safety."

IAEA Safety Standards Series No. GS-R-3, The Management System for Facilities and Activities [2], establishes the safety requirements for this quality assurance programme.

The management system is a set of interrelated or interacting elements that establishes policies and objectives and which enables those objectives to be achieved in a safe, efficient and effective manner. It integrates all aspects of managing a nuclear facility such as a research reactor, incorporating safety, health, quality, environment, security[1] and economic elements into one coherent management system. Such a management system is often referred to as an integrated management system.

The IAEA requirements and guidance use the term 'management system' rather than 'quality assurance'. The term 'management system' reflects and includes the initial concept of 'quality control' (controlling the quality of products) and its evolution through 'quality assurance' (the system to ensure the quality of products) and 'quality management' (the system to manage quality).

The publication Safety Guide Series No. GS-G-3.1 on the Application of the Management System for Facilities and Activities [3] provides generic guidance to aid the establishment, implementation, assessment and continual improvement of a management system that complies with the requirements laid out in in Ref. [2]. The Safety Guide Series No. GS-G-3.5, on The Management System for Nuclear Installations [4], provides specific guidance for research reactor operating organizations. The requirements presented in Ref. [2] and the guidance presented in Refs [3, 4] are applicable to research reactors, but a graded approach is needed to tailor the requirements and guidance for smaller facilities. The IAEA Safety Standard Series No. SSG-22 [5] provides detailed guidance on grading aspects for research reactors.

As established by Ref. [2], the operating organization is required to establish and implement a management system for the whole life cycle of a

---

[1] For operating organizations, security aspects are considered to be highly confidential and may not be included in the management system. However, their interaction with the management system processes ought to be known.

research reactor: siting, design, construction, commissioning, operation including utilization, modification, refurbishment and decommissioning.

For organizations operating low power research reactors with a limited number of experimental facilities, there are significant differences in the type and extent of the controls that need to be applied in comparison with those operating high power research reactors with an extended number of experimental facilities and an increased demand for isotope production or other irradiation services. Consequently, the scope, extent and details of the management system can be established and implemented using a graded approach. A supporting example on the use of the graded approach can be found in Ref. [5] and Annex II of this publication.

The development and implementation of a management system is a basic requirement in order to ensure:

— The safety of research reactors in all stages of their lifetime in order to protect the workers, the public and the environment from undue radiation hazards;
— Compliance with regulatory requirements;
— Proper and safe modification, refurbishment and upgrading;
— Safe and effective utilization of the research reactor for a variety of experiments and training as well as for commercial services.

Research reactors are used for specific and varying purposes, e.g. research, training, multi-elemental analysis, radioisotope production, neutron scattering, material irradiations, etc., resulting in different design features and operational regimes. Design and operating characteristics may vary significantly since experimental devices may affect the performance of research reactors.

The primary function of a research reactor is to support an effective and safe utilization programme. The reactor users are frequently from outside the operating organization or outside the normal reactor management line of control. Special arrangements are to be established for the proper lines of authority, communication and training of users in the safe design, construction, installation, operation, utilization and modification of experimental facilities [1, 6, 7].

1.2. OBJECTIVES

This Safety Report provides practical information and examples for the implementation of a management system for operating organizations of research reactors. This guidance conforms to the requirements and recommendations of the IAEA safety standards.

1.3. SCOPE

In the context of this publication, the term 'research reactor operating organization' covers the organization operating the reactor installation itself, its experimental facilities and all other facilities relevant to either the reactor or its experimental facilities located on the reactor site, including their supporting functions. This publication is applicable to operating research reactors and includes their utilization, modification, refurbishment and upgrading. With some slight modifications, this publication can also be used for research reactors in extended shutdown. Siting, design, construction, commissioning and decommissioning are other stages of a research reactor's lifetime with partly different processes. To tailor the management system to an actual situation, applicable processes have to be defined and developed for that particular stage.

1.4. STRUCTURE

Following this introductory section, Section 2 gives an overview of the prerequisites at the senior management level to beginning the implementation of a management system. Section 3 provides examples of the processes related to management responsibilities, while Sections 4 and 5 provide examples of the processes for resource management and process implementation, respectively. Section 6 provides examples of the processes for continual improvement of the management system. These processes are of a generic type and can be adapted where applicable for each operating organization depending on its size, organization and utilization. Section 7 provides a methodology to aid the development and implementation of a management system. Please note that the figures are grouped at the end of each section.

Figure 1 illustrates a generic operating organization structure with a three level management structure. In Sections 2–7, reference is made to these levels, as well as to the generic functions given in the different levels. The names of these functions will vary in different Member States and within operating organizations (e.g. the 'head' of operating organization may also be called 'director', 'CEO', 'president', etc.). The processes of the management system presented in Sections 3–6 and the implementation methodology given in Section 7 can easily be modified to reflect the nomenclature of different operating organizations. The annexes present practical examples of implementation in some organizations.

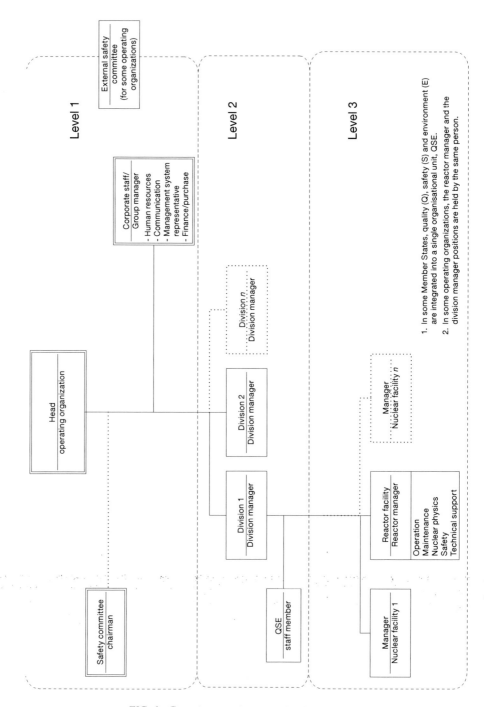

FIG. 1. Generic operating organization structure.

2. MANAGEMENT SYSTEM

2.1. ESTABLISHMENT AND IMPLEMENTATION OF A MANAGEMENT SYSTEM

The Management System for Facilities and Activities [2] requires that:

"A management system shall be established, implemented, assessed and continually improved. It shall be aligned with the goals of the organization and shall contribute to their achievement. The main aim of the management system shall be to achieve and enhance safety by:

— Bringing together in a coherent manner all the requirements for managing the organization;
— Describing the planned and systematic actions necessary to provide adequate confidence that all these requirements are satisfied;
— Ensuring that health, environment, security, quality and economic requirements are not considered separately from safety requirements, to help preclude their possible negative impact on safety.

Safety shall be paramount within the management system, overriding all other demands."

The following guidance has been developed to provide assistance in implementing this requirement for research reactors. It is supplementary to, and is to be read in conjunction with, the generic guidance established in Ref. [3].

— The goals of a research reactor apply to each division, unit, section and individual within the organization.
— A clear understanding of the sharing of responsibilities among all organizational units participating in or supporting the management system is necessary. These include centralized organizational and technical departments providing support, the operating organization's safety committees and any public services such as fire and medical services.

— The person in the senior management position (i.e. the head of the operating organization) in the organization is responsible for ensuring that the management system[2] is implemented.
— The establishment and implementation of the management system requires a combined effort on the part of managers, those performing the work and those assessing the work. Satisfactory implementation requires the planning and deployment of adequate resources.
— All staff are to be trained to achieve proficiency in and understanding of the management system processes that apply to the performance of their work. Management system effectiveness is assessed and reviewed at all stages of implementation. The information gained from assessments is used to achieve continuing improvements in work performance.

To establish a management system, an organization needs to:

— Review the applicable regulations and standards as well as the organization's management and technical practices to determine whether activities are adequately addressed;
— Review the applicable IAEA requirements such as Refs [1, 2], the management system Safety Guides [3, 4] and other relevant Safety Guides and Technical Reports to identify shortcomings and assign priorities to those areas requiring improvement or development;
— Establish timescales within which the required changes are to be implemented;
— Identify plans and document the activities that have to be carried out.

Senior management (levels 1 and 2 of Fig. 1) will ideally prepare a plan to establish, implement, evaluate and continually improve the management system. This plan must be communicated to all staff involved and must be implemented within a time frame that is commensurate with the activities it covers. The implementation plan has to be be approved and monitored by senior management.

It is recommended to implement the system while the implementation plan is being developed rather than waiting for the final plan to be completed. More detailed guidance on how to implement a management system is presented in Section 7 of this publication.

---

[2] The term 'management system' reflects and includes the concepts of quality control, quality assurance and quality management. The management system is a set of interrelated or interacting elements that establishes policies and objectives and which enables those objectives to be achieved in a safe, efficient and effective manner.

Staffing plans have to include provisions for selecting, training and assigning adequate numbers of personnel consistent with workload and schedules for implementation. Plans for assessing the effectiveness of processes and instructions and their implementation relative to the performance of work and the results achieved have to be specified. Frequent early assessments may be necessary to ensure the adequacy of instructions and to prevent the endorsement of poor practices.

2.2. SAFETY CULTURE

Safety requirements established in The Management System for Facilities and Activities [2] state that:

"The management system shall be used to promote and support a strong safety culture by:

— Ensuring a common understanding of the key aspects of safety culture within the organization;
— Providing the means by which the organization supports individuals and teams in carrying out their tasks safely and successfully, taking into account the interaction between individuals, technology and the organization;
— Reinforcing a learning and questioning attitude at all levels of the organization;
— Providing the means by which the organization continually seeks to develop and improve its safety culture."

It is recognized that all employees of the operating organization of a research reactor and all staff performing activities at this facility, such as researchers, students, fellows and scientists, contribute to the safety culture of the operating organization.

The management system must identify and address the different cultural aspects and behaviours of all staff of the operating organization. If there are any differences between working groups, then these differences have to be identified and action taken to harmonize them.

2.3. GRADING THE APPLICATION OF MANAGEMENT SYSTEM REQUIREMENTS

The Management System for Facilities and Activities [2] requires that:

"The application of management system requirements shall be graded so as to deploy appropriate resources, on the basis of the consideration of:

— The significance and complexity of each product or activity;
— The hazards and the magnitude of the potential impact (risks) associated with the safety, health, environmental, security, quality and economic elements of each product and activity;
— The possible consequences if a product fails or an activity is carried out incorrectly.

Grading of the application of management system requirements shall be applied to the products and activities of each process."

The following guidance has been developed to provide a means of implementing this requirement for research reactors. It is supplementary to, and is to be read in conjunction with, the generic guidance established in Ref. [3].

Although the establishment and implementation of a management system is a requirement for an operating organization of a research reactor, the application of the requirements and controls of the management system can be performed using a graded approach. The grading can be based on the safety relevance of specific processes and activities, or to minimize risk. All the requirements of the management system as defined in Ref. [2] are to be applied unless it can be justified that, for the research reactor in question, the application of certain requirements can be graded. For each case, the application of the requirements to be graded has to be identified, with account taken of the nature and possible magnitude of the hazards presented by the research reactor and the activities conducted [1, 5].

It would be beneficial to develop and apply a graded approach as early as possible. The established list of components, structures and systems with specific relevance to safety is the starting point for any grading. Special attention is to be given to the utilization of a research reactor, e.g. the execution of irradiation experiments and scientific research where the grading of requirements could be performed on the basis of engineering judgment. A typical example is the more stringent licensing, inspection and audit requirements for prototype fuel elements in comparison to standard fuel elements. Guidance on the appropriate use of grading is given in Ref. [5] while a typical example of a graded approach can be found in Annex I.

2.4. DOCUMENTATION OF THE MANAGEMENT SYSTEM

2.4.1. General

The Management System for Facilities and Activities [2] requires that:

"The documentation of the management system shall include the following:

— The policy statements of the organization;
— A description of the management system;
— A description of the structure of the organization;
— A description of the functional responsibilities, accountabilities, levels of authority and interactions of those managing, performing and assessing work;
— A description of the processes and supporting information that explain how work is to be prepared, reviewed, carried out, recorded, assessed and improved.

The documentation of the management system shall be developed to be understandable to those who use it. Documents shall be readable, readily identifiable and available at the point of use.
The documentation of the management system shall reflect:

— The characteristics of the organization and its activities;
— The complexities of processes and their interactions."

The following guidance has been developed to provide assistance in implementing this requirement for research reactors. It is supplementary to, and is to be read in conjunction with, the generic guidance established in Ref. [3].

It is beneficial if the content of management system documents is determined in consultation with managers and staff who will use them to do their work, and others in the operating organization who might be affected by them. Such individuals will ideally also have an input into subsequent revisions. In the case of detailed working documents, trial use and validation using mock-ups, simulators and walk-throughs, or preproduction runs and testing, are ways of determining the accuracy of the documents.

2.4.2. Information structure

The IAEA Safety Guide on the Application of the Management System for Facilities and Activities [3] provides the following guidance:

"A three level structure of information promotes clarity and avoids repetition by establishing the amount of information and the level of detail appropriate to each type of document and by using cross-references between specific documents at the different levels. A typical three level structure consists of:
— Level 1 — An overview of how the organization and its management system are designed to meet its policies and objectives;
— Level 2 — A description of the processes to be implemented to achieve the policies and objectives and the specification of which organizational unit is to carry them out;
— Level 3 — Detailed instructions and guidance that enable the processes to be carried out and specification of the individual or unit that is to perform the work."

Some research establishments may be complex operating organizations with facilities at several sites, or with a large department with radiation related activities, e.g. a radiotherapy department in a hospital. In such cases, it is recommended to establish common level 1 and 2 documentation of the management system integrating the general objectives, mission and work of the operating organization. To complement such a management system, specific working level documents (level 3) are generally necessary to address work that is unique to one or more of the organization's facilities or sites.

The number and details of the processes within a management system also has to be based on the safety relevance of the systems or activities to be performed.

3. MANAGEMENT RESPONSIBILITY

3.1. MANAGEMENT COMMITMENT

The Management System for Facilities and Activities [2] requires that:

"Management at all levels shall demonstrate its commitment to the establishment, implementation, assessment and continual improvement of the management system and shall allocate adequate resources to carry out these activities.

"Senior management shall develop individual values, institutional values and behavioural expectations for the organization to support the implementation of the management system and shall act as role models in the promulgation of these values and expectations.

"Management at all levels shall communicate to individuals the need to adopt these individual values, institutional values and behavioural expectations as well as to comply with the requirements of the management system.

"Management at all levels shall foster the involvement of all individuals in the implementation and continual improvement of the management system.

"Senior management shall ensure that it is clear when, how and by whom decisions are to be made within the management system."

The guidance in this publication has been developed to provide a means of implementing this requirement for research reactors. It is supplementary to, and is to be read in conjunction with, the generic guidance established in Ref. [3].

Senior managers have to empower individuals and make them feel responsible for their work. In this way, senior management can encourage improved individual and organizational performance.

Line managers, supervisors and team leaders[3] (see Fig. 1) are responsible for ensuring that personnel working under their supervision have been provided with the necessary training, resources and directions. These elements have to be provided before any work begins. It is recognized that within operating organizations the complexity of the different processes and span of control of the line managers, supervisors and team leaders can vary.

---

[3] Line managers, supervisors and team leaders are those persons within the operating organization with direct responsibility for and supervision of a group of employees or of one or more processes. They can belong either to level 2 or level 3 of the organization shown in Fig. 1.

Line managers, supervisors and team leaders have to examine a sample of work practices and related information on a regular basis to ensure that the expected level of performance is being achieved and to identify areas that need to be improved. They have to also encourage each individual under their supervision to look for more efficient and effective ways of accomplishing assigned tasks.

3.2. MANAGEMENT PROCESSES

3.2.1. General

The following management processes are provided as an example to illustrate a typical set of processes that could be used by a research reactor operating organization. However, each research reactor operating organization has to apply the guidance provided in this publication to the development of their own processes. The legend, acronyms and instructions for use of the following sample processes can be found following the definitions.

3.2.2. Planning and reporting[4]

The process shown in Fig. 2 describes a methodology used to establish, review and authorize long term operational plans[5] at both organizational and divisional levels, as well as financial reporting.

3.2.3. Management system administration

The process shown in Fig. 3 describes the methodology used to issue a new or revised process at the organization and divisional level. It also shows progress monitoring, review and authorization as well as the controlled distribution. The term QSE is an abbreviation of quality, safety and environment.

3.2.4. Organizational structure

The process shown in Fig. 4 describes a methodology used to identify or adapt the structure of the operating organization as well as substitution schemes and mandates at organizational and divisional level. It also provides for the introduction and authorization of (new) job descriptions.

---

[4] Planning and reporting includes financial aspects.
[5] For some operating organizations, the mission and policy are fixed and cannot be changed for legal reasons. In those cases, the input indicated going to step 1 goes instead to step 2.

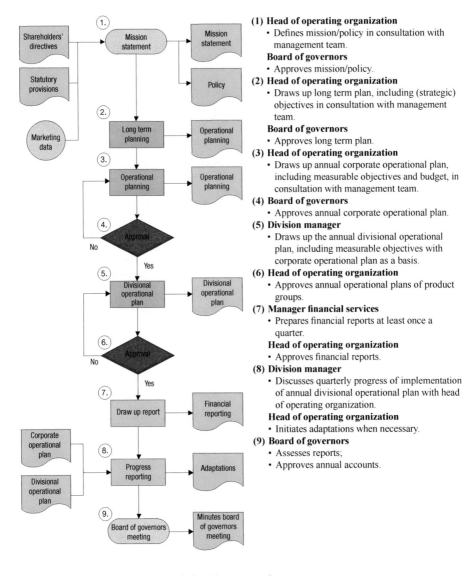

(1) **Head of operating organization**
  • Defines mission/policy in consultation with management team.
  Board of governors
  • Approves mission/policy.

(2) **Head of operating organization**
  • Draws up long term plan, including (strategic) objectives in consultation with management team.
  Board of governors
  • Approves long term plan.

(3) **Head of operating organization**
  • Draws up annual corporate operational plan, including measurable objectives and budget, in consultation with management team.

(4) **Board of governors**
  • Approves annual corporate operational plan.

(5) **Division manager**
  • Draws up the annual divisional operational plan, including measurable objectives with corporate operational plan as a basis.

(6) **Head of operating organization**
  • Approves annual operational plans of product groups.

(7) **Manager financial services**
  • Prepares financial reports at least once a quarter.
  Head of operating organization
  • Approves financial reports.

(8) **Division manager**
  • Discusses quarterly progress of implementation of annual divisional operational plan with head of operating organization.
  Head of operating organization
  • Initiates adaptations when necessary.

(9) **Board of governors**
  • Assesses reports;
  • Approves annual accounts.

FIG. 2. Planning and reporting.

13

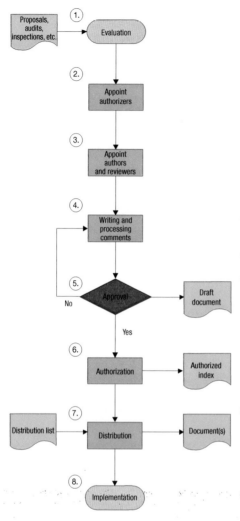

(1) **Management system representative/QSE staff member**
  - Evaluates the necessity for revision at least once every 3 years;
  - Evaluates proposals for organization or group specific documents.

(2) **Management system representative**
  - Decides in consultancy with QSE staff member(s) whether a corporate document needs to be drafted or withdrawn;
  - Appoints authorizer for organization specific documents.

  Division manager
  - Decides whether a division specific document needs to be drafted or withdrawn;
  - Appoints authorizer for division specific documents.

(3) **Authorizer**
  - Appoints author(s) and reviewers (for corporate documents the QSE committee is the reviewer).

(4) **Author**
  - Prepares draft and processes comments.

(5) **Reviewer(s)**
  - Reviews draft documents.

  Management system representative
  - Monitors progress of corporate documents and decides in consultancy with QSE staff members on implementation;
  - Management team decides on implementation in case of disagreement within QSE committee.

  Authorizer
  - Monitors progress and implementation of decisions.

(6) **Management system representative/QSE staff member**
  - Ensures correct layout;
  - Indicates modifications from the previous version in index.

  Head of operating organization
  - Authorizes procedures.

  Management system representative
  - Authorizes supporting documents.

  Authorizer division specific documents
  - Authorizes division specific documents.

(7) **Management system representative/QSE staff member**
  - Distributes documents and supervises archiving of original and expired version(s).

  QSE staff member
  - Sends a copy of the local division specific index to management system representative.

(8) **Division manager**
  - Ensures implementation.

FIG. 3. Management system administration.

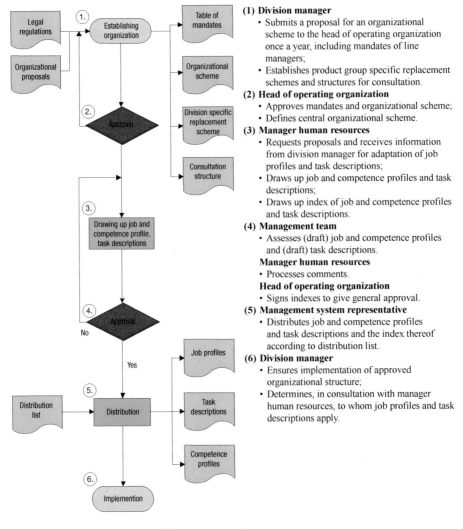

(1) Division manager
- Submits a proposal for an organizational scheme to the head of operating organization once a year, including mandates of line managers;
- Establishes product group specific replacement schemes and structures for consultation.

(2) Head of operating organization
- Approves mandates and organizational scheme;
- Defines central organizational scheme.

(3) Manager human resources
- Requests proposals and receives information from division manager for adaptation of job profiles and task descriptions;
- Draws up job and competence profiles and task descriptions;
- Draws up index of job and competence profiles and task descriptions.

(4) Management team
- Assesses (draft) job and competence profiles and (draft) task descriptions.

Manager human resources
- Processes comments.

Head of operating organization
- Signs indexes to give general approval.

(5) Management system representative
- Distributes job and competence profiles and task descriptions and the index thereof according to distribution list.

(6) Division manager
- Ensures implementation of approved organizational structure;
- Determines, in consultation with manager human resources, to whom job profiles and task descriptions apply.

FIG. 4. Organizational structure.

4. RESOURCE MANAGEMENT

The Management System for Facilities and Activities [2] requires that:

"Senior management shall determine the amount of resources necessary and shall provide the resources[9] to carry out the activities of the organization and to establish, implement, assess and continually improve the management system.

"The information and knowledge of the organization shall be managed as a resource.

"Senior management shall determine the competence requirements for individuals at all levels and shall provide training or take other actions to achieve the required level of competence. An evaluation of the effectiveness of the actions taken shall be conducted. Suitable proficiency shall be achieved and maintained.

"Senior management shall ensure that individuals are competent to perform their assigned work and that they understand the consequences for safety of their activities. Individuals shall have received appropriate education and training, and shall have acquired suitable skills, knowledge and experience to ensure their competence. Training shall ensure that individuals are aware of the relevance and importance of their activities and of how their activities contribute to safety in the achievement of the organization's objectives.

"Senior management shall determine, provide, maintain and re-evaluate the infrastructure and the working environment necessary for work to be carried out in a safe manner and for requirements to be met.

[9] 'Resources' includes individuals, infrastructure, the working environment, information and knowledge, and suppliers, as well as material and financial resources."

The following guidance has been developed to provide a means of implementing these requirements for research reactors. It is supplementary to, and is to be read in conjunction with, the generic guidance established in Ref. [3].

The following resource related processes are provided as an example to illustrate a typical set of processes that could be used by a research reactor organization. The management of financial aspects is seen as a management process and is covered in Section 3.2.2, 'Planning and reporting'.

4.1. RECRUITMENT, SELECTION AND APPOINTMENT

The process shown in Fig. 5 describes a methodology used to define job vacancies and to manage the selection and recruitment of new employees. It also includes the initial safety related training requirements at the organizational and divisional level. For further information see Ref. [6].

4.2. PERFORMANCE, ASSESSMENT AND TRAINING

The process shown in Fig. 6 describes a methodology used to check and improve the performance of all employees of an operating organization. By means of (annual) interviews and appraisal talks the need for (re)training and job rotation is also investigated. Furthermore, the documentation of results and records is included. For further information see Ref. [6].

4.3. EMPLOYING TEMPORARY WORKERS OR OUTSOURCING

The process shown in Fig. 7 describes a methodology applied to a job vacancy which will be filled by a temporary worker or for which activities are outsourced. It includes the tasks and responsibilities of the various staff involved (line management, human resources, purchase department, etc.) and a check for compliance with regulatory and legal requirements. In the examples, the term outsourcing is used to describe contracting out work. When temporary workers are hired, they are considered part of the operating organization for a limited period of time.

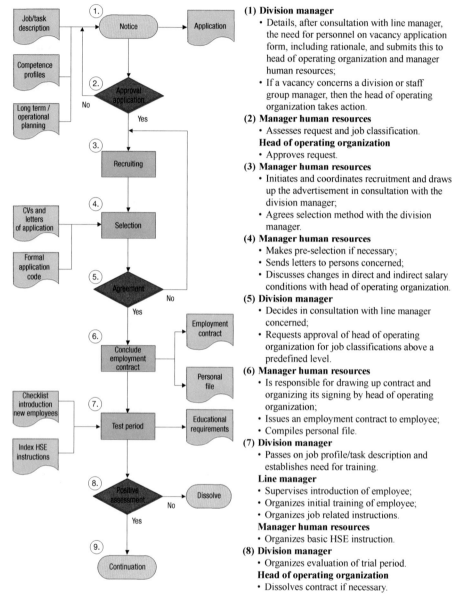

(1) Division manager	
• Details, after consultation with line manager, the need for personnel on vacancy application form, including rationale, and submits this to head of operating organization and manager human resources;	
• If a vacancy concerns a division or staff group manager, then the head of operating organization takes action.	

(1) Division manager
- Details, after consultation with line manager, the need for personnel on vacancy application form, including rationale, and submits this to head of operating organization and manager human resources;
- If a vacancy concerns a division or staff group manager, then the head of operating organization takes action.

(2) Manager human resources
- Assesses request and job classification.

Head of operating organization
- Approves request.

(3) Manager human resources
- Initiates and coordinates recruitment and draws up the advertisement in consultation with the division manager;
- Agrees selection method with the division manager.

(4) Manager human resources
- Makes pre-selection if necessary;
- Sends letters to persons concerned;
- Discusses changes in direct and indirect salary conditions with head of operating organization.

(5) Division manager
- Decides in consultation with line manager concerned;
- Requests approval of head of operating organization for job classifications above a predefined level.

(6) Manager human resources
- Is responsible for drawing up contract and organizing its signing by head of operating organization;
- Issues an employment contract to employee;
- Compiles personal file.

(7) Division manager
- Passes on job profile/task description and establishes need for training.

Line manager
- Supervises introduction of employee;
- Organizes initial training of employee;
- Organizes job related instructions.

Manager human resources
- Organizes basic HSE instruction.

(8) Division manager
- Organizes evaluation of trial period.

Head of operating organization
- Dissolves contract if necessary.

FIG. 5. Recruitment, selection and appointment.

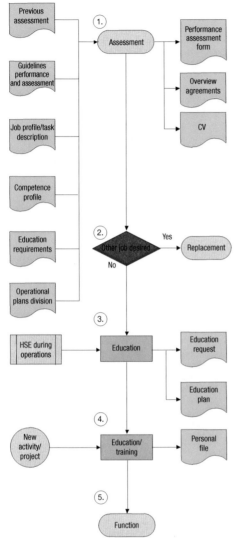

(1) Division manager
- Coordinates performance review (if a division manager is concerned, the head of operating organization takes action).

Division manager/line manager
- Organizes interviews;
- Defines education/training requirements;
- Defines qualifications for specific assignments if necessary;
- Adapts CV if necessary;
- Signs off on performance/assessment form.

Division manager
- Endorses assessment.

Manager human resources
- Adds form and data to personal file;
- Sends overview of actions of performance/assessment interview to division manager.

(2) Division manager
- Initiates reappointment if necessary.

Manager human resources
- Checks and advises head of operating organization.

Head of operating organization
- Decides on reappointment.

Manager human resources
- Adds data to personal file.

(3) Division manager/line manager
- Determines if additional education/training is required;
- Prepares education/training plan;
- Appoints employee for HSE instruction.

Division manager
- Approves education/training plan;
- Sends application form to manager human resources.

Employee
- Requests additional education/training provisions if desired.

(4) Division manager/line manager
- Provides resources to implement education/training;
- Determines whether training plan has been completed and adapts plan if required;
- Checks employee's qualification if necessary;
- Sends education/training results to manager human resources.

Manager human resources
- Adds education/training data to personal file.

FIG. 6. Performance, assessment and training.

19

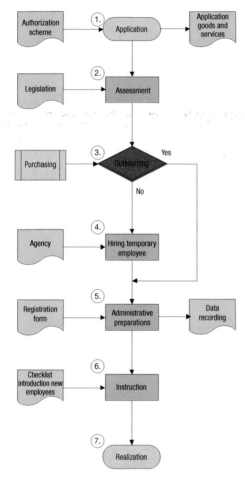

(1) **Line manager**
- Fills out application form goods and services, including qualification, period, job level;
- Organizes authorization in conformance with authorization scheme.

Authorizer(s)
- Give(s) approval.

(2) **Manager human resources**
- Assesses desirability;
- Checks legal regulations;
- Organizes formal approval by managing director;
- Sends application form to financial services.

(3) **Purchaser**
- Coordinates completion — in case of outsourcing, according to the procedure in Section 5.1.5.

(4) **Division manager**
- Consults with financial services and human resources when a new organization is involved;
- Coordinates selection of employee in consultation with applicant;
- Sends application form to financial services if outsourcing is necessary.

(5) **Division manager**
- Organizes the provision of personal data to human resources.

Manager human resources
- Supervises the recording of personal data;
- Supervises filing, including qualifications;
- Provides data to financial services.

Financial services
- Inputs data into business information system.

(6) **Manager human resources**
- Supplies checklist 'introduction of new employees';
- Organizes basic HSE instruction.

Line manager
- Coordinates workplace related instructions;
- Gives instructions for time registration.

(7) **Line manager**
- Authorizes time sheets;
- Sends copy to financial services;
- Notifies end of contract to financial services and human resources.

Financial services/human resources
- Dissolves contract administratively.

FIG. 7. Employing temporary workers or outsourcing.

20

5. PROCESS IMPLEMENTATION

5.1. GENERIC MANAGEMENT SYSTEM PROCESSES

The Management System for Facilities and Activities [2] requires the development of the following generic processes in the management system.

5.1.1. Control of documents

Ref. [2] states that:

"Documents[10] shall be controlled. All individuals involved in preparing, revising, reviewing or approving documents shall be specifically assigned this work, shall be competent to carry it out and shall be given access to appropriate information on which to base their input or decisions. It shall be ensured that document users are aware of and use appropriate and correct documents.

"Changes to documents shall be reviewed and recorded and shall be subject to the same level of approval as the documents themselves.

[10] Documents may include: policies; procedures; instructions; specifications and drawings (or representations in other media); training materials; and any other texts that describe processes, specify requirements or establish product specifications."

The generic guidance that has been developed to provide a means of implementing this requirement is established in Ref. [3].

5.1.2. Control of products

Ref. [2] requires:

"Specifications and requirements for products, including any subsequent changes, shall be in accordance with established standards and shall incorporate applicable requirements. Products that interface or interact with each other shall be identified and controlled.

"Activities for inspection, testing, verification and validation shall be completed before the acceptance, implementation or operational use of products. The tools and equipment used for these activities shall be of the proper range, type, accuracy and precision.

21

"The organization shall confirm that products meet the specified requirements and shall ensure that products perform satisfactorily in service.

"Products shall be provided in such a form that it can be verified that they satisfy the requirements.

"Controls shall be used to ensure that products do not bypass the required verification activities.

"Products shall be identified to ensure their proper use. Where traceability is a requirement, the organization shall control and record the unique identification of the product.

"Products shall be handled, transported, stored, maintained and operated as specified, to prevent their damage, loss, deterioration or inadvertent use."

5.1.3. Measuring and test equipment

The following guidance has been developed to provide a means of implementing the above mentioned requirements related to activities involved in the inspection, testing, verification and validation of research reactors. It is supplementary to, and is to be read in conjunction with the generic guidance established in Ref. [3].

A process for the control, and, where necessary, calibration of tools, gauges, instruments and other measuring, inspecting and testing equipment used for activities important to the safety of the facility have to be established.

A process for the monitoring of out of calibration equipment, including segregation to prevent further use and identification and evaluation of the impact of the use on previous measurements since the last calibration date, has to be established.

Tools, gauges, instruments and other measuring, inspecting and testing equipment (including test software and devices) used in determining product status and verifying the acceptability of products have to be of the proper range, type, accuracy and measuring precision.

The selection, identification, use, calibration requirements and calibration frequency of all measuring, inspecting and testing equipment used for the determination of product quality or operational status has to be specified. The responsibility for measuring and for test equipment controls has to be defined. Such aspects include:

— Identification of the measurements to be made, the accuracy required and the selection of the appropriate measuring and test equipment.

— Identification, calibration and adjustment of all measuring and test equipment and devices that can affect product quality at prescribed intervals, or prior to use, against certified equipment having a known and valid relationship to nationally or internationally recognized standards. If such standards do not exist, the basis used for calibration has to be documented.
— Establishment, documentation and maintenance of calibration procedures that include details of equipment type, unique identification number, location, frequency of checks, check method, acceptance criteria and the action to be taken when results are unsatisfactory.
— Confirmation that the measuring and test equipment is capable of the accuracy and precision necessary.
— Identification of measuring and test equipment with a suitable indicator or approved identification record to show its calibration status.
— Maintenance of calibration records for measuring and test equipment.
— The review and documenting of the validity of previous measurements when measuring and test equipment is found to be out of calibration. Operation and maintenance records can be used as a source of information to identify which test equipment was used.
— Controls to ensure that environmental conditions are suitable for the calibration, measurements and tests being carried out.
— Controls to ensure that the handling, preservation, storage and use of calibrated equipment is such that its accuracy and fitness for use is maintained.
— Protecting measuring and test equipment from adjustments which may invalidate their accuracy.
— Methods for adding and removing measuring and test equipment to and from the calibration programme, including the means to ensure that new or repaired products are calibrated prior to their use.
— A process to control the issue of measuring and test equipment to qualified and authorized individuals.
— Testing hardware, such as jigs, fixtures, templates or patterns, and/or test software used for inspection, have to be checked prior to use in production and within the facility, and rechecked at prescribed intervals. The extent and frequency of such checks has to be established and records maintained as evidence of control. Such approved testing hardware has to be properly identified.

The sample process shown in Fig. 8 describes the methodology for calibration, verification and documentation of pre-indicated measuring devices. It also includes the handling of non-conformances and withdrawal from or release for use of the devices. It is supplementary to, and is to be read in conjunction with, the generic guidance established in Ref. [3].

5.1.4. Control of records

The Management System for Facilities and Activities [2] requires:

"Records shall be specified in the process documentation and shall be controlled. All records shall be readable, complete, identifiable and easily retrievable.

"Retention times of records and associated test materials and specimens shall be established to be consistent with the statutory requirements and knowledge management obligations of the organization. The media used for records shall be such as to ensure that the records are readable for the duration of the retention times specified for each record."

The generic guidance that has been developed to provide a means of implementing this requirement is established in Ref. [2].

5.1.5. Purchasing

Ref. [2] requires that:

"Suppliers of products shall be selected on the basis of specified criteria and their performance shall be evaluated.

"Purchasing requirements shall be developed and specified in procurement documents. Evidence that products meet these requirements shall be available to the organization before the product is used.

"Requirements for the reporting and resolution of non-conformances shall be specified in procurement documents."

The following sample process has been developed to provide a means of implementing this requirement for research reactors. It is supplementary to, and is to be read in conjunction with the generic guidance established in Ref. [3].

The process shown in Fig. 9 describes the methodology for the purchase and receipt of goods including the selection of (preferred) suppliers. It also includes the monitoring of agreed delivery times and recording of possible deviations.

5.1.6. Communication

The Management System for Facilities and Activities [2] requires that:

"Information relevant to safety, health, environmental, security, quality and economic goals shall be communicated to individuals in the organization and, where necessary, to other interested parties.

"Internal communication concerning the implementation and effectiveness of the management system shall take place between the various levels and functions of the organization."

The generic guidance that has been developed to provide a means of implementing this requirement is established in Ref. [3].

5.1.7. Managing organizational change

The Management System for Facilities and Activities [2] requires that: "Organizational changes shall be evaluated and classified according to their importance to safety and each change shall be justified. The implementation of such changes shall be planned, controlled, communicated, monitored, tracked and recorded to ensure that safety is not compromised."

More guidance that has been developed to provide a means of implementing this requirement is established in Ref. [3]. More guidance on safety aspects to be considered in managing organizational change is provided in Ref. [8]. A sample process for managing safety related organizational changes is shown in Fig. 10.

5.2. PROCESSES COMMON TO ALL RESEARCH REACTORS

The following sample processes are common to all research reactors. The guidance has to be utilized taking into account the life cycle of the research reactor, the organization's size and structure and the nature of the activities to be carried out.

5.2.1. Project management

The process shown in Fig. 11 describes a methodology for appointing a project leader and for setting up a project including its financial, technical and planning aspects, and for submitting the results to the customer. It also requires the assessment of whether a project evaluation is needed and its execution if necessary.

This process can also be used for projects to modify the installation and/or equipment using the procedure for modifications of installations. The procedure for modification of installations is installation specific. Normally, the review and approval routes are based on the safety classification of the installation and the configuration management aspects are part of this procedure. An example of such aspects, which have to be considered in the configuration management process, is given in Annex III. More information on configuration management can be found in Refs [9, 10].

5.2.2. Quotation qualification

The process shown in Fig. 12 describes a methodology used to prepare, authorize, issue and register potential external customers. It also gives a process for comparing the initial qualification with the actual order and beginning the execution of any contracted work.

5.2.3. Preparation of technical reports

The process shown in Fig. 13 describes a methodology for writing, reviewing and authorizing reports from the operating organization. It also includes an independent review to check consistency with the organization's house style, as well as copying, distributing and archiving.

5.2.4. Numerical calculations

The process shown in Fig. 14 describes a methodology for selecting codes, standards and modelling requirements when numerical calculations are performed. It also includes the review of input data and the independent verification of results.

5.2.5. Incoming and outgoing correspondence

The process shown in Fig. 15 describes a methodology for the receipt, forwarding, distribution and filing of correspondence, both incoming (i.e. received from external parties) and outgoing (i.e. sent to external parties). It includes, among other instructions, requirements for any correspondence related to licences.

5.2.6. Archiving

The process shown in Fig. 16 describes a methodology for identifying and maintaining an archive and term list, including the expiration date of documents to be archived. It also includes the tasks of the staff member(s) responsible for archiving.

5.2.7. Software administration

The process shown in Fig. 17 describes a methodology used for software administration and to determine the responsibilities of the staff involved. It also includes testing, backups of the software and reporting of (potential) malfunctions.

5.2.8. Administration and maintenance of equipment and installations

The process shown in Fig. 18 describes a methodology used to determine and implement periodic testing and preventive maintenance programmes. It also includes feedback from users and/or operators of the facility and the recording/archiving of the results. In this example ageing management is regarded as a part of the maintenance programme, but it can also be seen as a separate subprocess. An example of ageing management aspects to be considered is presented in Annex IV. More safety related guidance is provided in Ref. [11].

5.2.9. Operation of installations and laboratories

The process shown in Fig. 19 describes a methodology used to safely operate nuclear installations and laboratories according to the annual operational plan and within the operating limits and conditions of applicable licenses or authorizations. It also includes the need to prepare a technical information package[6], a preventive maintenance programme and health, safety and environmental instructions. Furthermore, it includes the review and authorization of the aforementioned documents.

The technical information package in this example includes:

— A detailed description of the facility;
— Safety analysis report;

---

[6] For some operating organizations, the technical information package is included in the safety analysis report.

— As-built drawings;
— Design data;
— Identification of special requirements (radiation protection, fire protection, isolation and tagging, inspection and testing etc.);
— Environmental impact statement.

More safety related guidance is provided in Refs [7, 9].

5.2.10. Waste management

The process shown in Fig. 20 describes a methodology used to handle all the waste streams (gaseous, liquid and solid) of an operating organization. It separates the effluents into radiological, toxic and inflammable waste etc. with dedicated management of each type. More safety related information is provided in Ref. [12].

5.2.11. Health, safety and environmental aspects of operations

The process shown in Fig. 21 describes a methodology used to implement the integration of health, safety and environmental (HSE) aspects in the daily operation of a nuclear facility. It also includes the content of the inspection and testing programmes as well as the frequency of periodic inspections and the requirements for toolbox[7] meetings. Furthermore, it indicates the need to incorporate HSE aspects, including (near) misses, in regular meetings.

5.2.12. Potentially unsafe situations

A potentially unsafe situation could be an abnormal event, an accident, a lack of a safety related procedure or a situation with a potential risk which could have been avoided. The process shown in Fig. 22 describes a methodology used to report, handle and resolve a potentially unsafe situation. It also includes the reporting route[8] to the competent authorities and ways to improve by implementing lessons learned in regular meetings.

---

[7] A toolbox meeting is a meeting on a periodic or as-needed basis in which the preparations and detailed aspects of specific activities are discussed between workers and their supervisor.

[8] Notifications are also sent to the reactor safety committee (RSC) for review at their periodic meetings.

5.2.13. Accident and/or emergency

The process shown in Fig. 23 describes a methodology used to manage emergency situations on the site of the operating organization. It also includes formal reporting[9] to the competent authorities as well as the tasks and responsibilities of the various staff involved. Detailed instructions, communication lines, up-scaling and required resources are specified in the dedicated emergency plans.

5.2.14. Reactor safety committee (RSC)

The process shown in Fig. 24 describes a methodology used to seek advice from the organization's reactor safety committee (RSC)[10]. It contains a summary of different items and activities for which RSC review is mandatory and of how RSC advice[11] is to be implemented into the operation of the nuclear facilities. More information is provided in Ref. [1].

5.2.15. Licence, occupational health and safety assessment

The process shown in Fig. 25 describes a methodology for applying, implementing or revising licence and/or occupational HSE aspects at the organizational level. More information is provided in Refs [12, 13].

Text cont. on p. 48.

---

[9] Notifications are also sent to the reactor safety committee (RSC) and the management system representative for review in their periodic meetings.

[10] Although not commonly present, an external safety committee can be useful for large operating organizations.

[11] The RSC meets on a regular basis and also as needed by the operating organization.

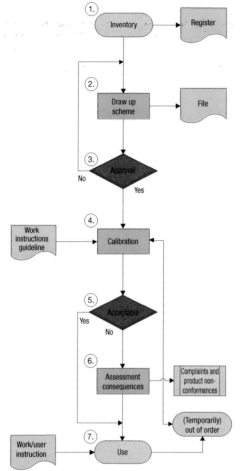

(1) Line manager
- Establishes register of measuring devices and reference materials to be calibrated or verified;
- Appoints calibration technician(s).

(2) Calibration technician
- Compiles or amends files;
- Includes in scheme, if applicable, interval, method, criteria for acceptance or adjustment and measurement uncertainty during use. It is periodically decided whether the interval and/or the measurement uncertainty remain the same.

(3) Line manager
- Approves method, criteria and scheme.

(4) Calibration technician
- Coordinates calibration and verification.

(5) Calibration technician
- Assesses the results;
- Identifies suspect measurement results;
- Analyses problem and proposes solution;
- Ensures that calibration date can be traced at installation.

(6) Line manager
- Checks consequences;
- Initiates complaints and non-conformance procedure for recall of results and/or informing clients;
- Is responsible for the implementation of corrective/preventive actions.

Calibration technician
- Clearly labels measurement device as non-operational.

(7) Calibration technician
- Makes measurement device(s) available to user;
- Withdraws measurement device from use if necessary.

User
- Uses measurement device according to instructions;
- Reports malfunctioning, problems or suspected disorders to calibration technician.

FIG. 8. Calibration of measuring devices.

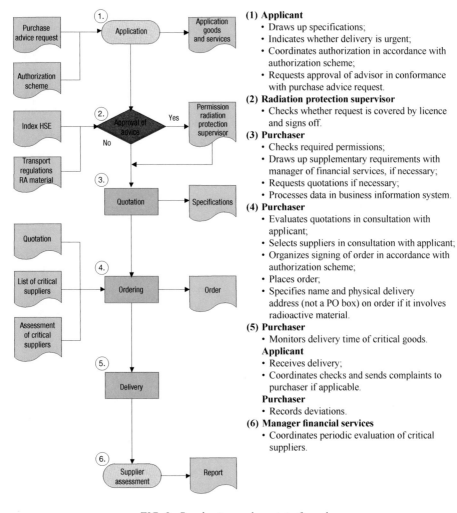

(1) Applicant
- Draws up specifications;
- Indicates whether delivery is urgent;
- Coordinates authorization in accordance with authorization scheme;
- Requests approval of advisor in conformance with purchase advice request.

(2) Radiation protection supervisor
- Checks whether request is covered by licence and signs off.

(3) Purchaser
- Checks required permissions;
- Draws up supplementary requirements with manager of financial services, if necessary;
- Requests quotations if necessary;
- Processes data in business information system.

(4) Purchaser
- Evaluates quotations in consultation with applicant;
- Selects suppliers in consultation with applicant;
- Organizes signing of order in accordance with authorization scheme;
- Places order;
- Specifies name and physical delivery address (not a PO box) on order if it involves radioactive material.

(5) Purchaser
- Monitors delivery time of critical goods.

Applicant
- Receives delivery;
- Coordinates checks and sends complaints to purchaser if applicable.

Purchaser
- Records deviations.

(6) Manager financial services
- Coordinates periodic evaluation of critical suppliers.

FIG. 9. Purchasing and receipt of goods.

31

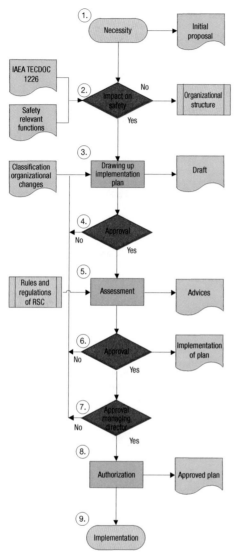

(1) **Responsible manager**
   - Identifies necessary organizational changes;
   - Draws up initial proposal.

(2) **Division manager**
   - Assesses proposal;
   - Appoints drafters and internal reviewers if the subject is safety related.

(3) **Drafter**
   - Draws up draft implementation plan;
   - Proposes classification;
   - Processes comments.

   Reviewers
   - Provides initial assessment.

(4) **Division manager**
   - Approves proposal;
   - Starts, if necessary, an RSC* assessment;
   - Requests advice, if necessary, from manager human resources, management system representative and radiation protection supervisor.

(5) **RSC***
   - Performs assessment and prepares advice.

   Manager human resources
   - Requests assessment from management team and staff council, if necessary;
   - Prepares advice.

   Radiation protection supervisor
   - Prepares advice.

   Management system representative
   - Prepares advice.

(6) **Division manager**
   - Assesses advices;
   - Adjusts, if necessary, the implementation plan;
   - Approves implementation plan.

   Head of operating organization
   - Decides between conflicting views and informs staff involved.

(7) **Head of operating organization**
   - Requests, if necessary, the approval of the regulatory body.

(8) **Head of operating organization**
   - Authorizes implementation plan.

(9) **Division manager**
   - Ensures that implementation takes place.

* RSC: Reactor safety committee

FIG. 10. Managing organizational changes.

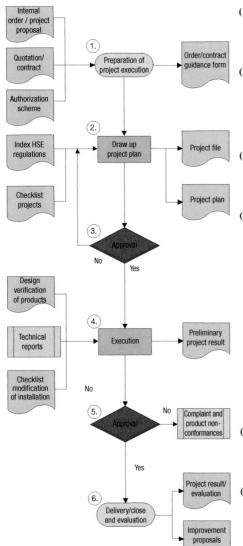

(1) **Line manager**
- Appoints project manager(s);
- Makes initial budget available as necessary;
- Organizes drawing up of order and contract guidance form.

(2) **Project manager**
- Opens project file;
- Draws up project plan in consultation with line manager involved (concerning finances, time schedule, organization, information, quality, resources, results, working procedures, HSE aspects and communication).

(3) **Division manager**
- Evaluates draft project plan in consultation with account manager;
- Approves project plan;
- Releases project budget.

(4) **Financial services**
- Organizes financial project file, including financial and administrative data.

Project manager
- Keeps project file up to date;
- Maintains technical and substantive contacts with customer (contacts on management level through account manager);
- Controls substantive and financial progress;
- Initiates invoicing;
- Coordinates design assessment in case of design activities;
- Reports to division manager and/or line manager with respect to progress, bottlenecks and non-conformances;
- Delivers provisional project results;
- Checks contract requirements and initiates contract amendments if necessary.

(5) **Division manager or line manager**
- Checks provisional project results;
- Initiates corrective actions in consultation with project manager or account manager in case of non-conformances.

(6) **Division or line manager**
- Delivers project results to customer;
- Initiates final settlement of accounts and closes if needed;
- Performs project evaluation in consultation with division manager;
- Provides relevant supplier information to financial services;
- Prepares project file for archiving.

Division manager
- Reviews project evaluation and proposes improvements, if necessary;
- Closes project.

Manager financial services
- Closes the financial and administrative files.

FIG. 11. Project management.

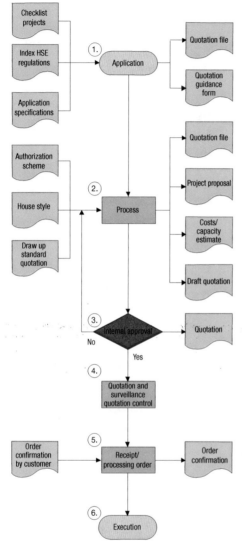

(1) Division manager
- Evaluates desirability of issuing quotation and hands over to other division manager if necessary;
- Appoints quotation manager;
- Organizes drawing up of quotation form;
- Assigns acquisition budget if desired;
- Informs applicant if request will not be taken forward;
- Informs manager financial services.

Manager financial services
- Supervises registration;
- Opens acquisition budget;
- Organizes financial quotation file.

(2) Quotation manager
- Opens quotation file;
- Draws up project proposal and technical specification, including internal cost estimate, resource plan, risk assessment and time schedule in consultation with relevant line manager and, if necessary, the account manager;
- Draws up quotation.

Manager financial services
- Evaluates quotations with high risks.

(3) Quotation manager
- Coordinates signing and distribution of quotation form;
- Coordinates signing of quotation in accordance with authorization scheme;
- Coordinates submission and distribution.

(4) Quotation manager
- Monitors progress of quotation;
- Maintains contact with applicant;
- Applies procedure in case of an amendment;
- Informs those involved if order is cancelled;
- Concludes acquisition.

(5) Quotation manager
- Evaluates whether order is in accordance with quotation;
- Formulates order confirmation and contract in consultation with manager financial services and establishes deviations;
- Requests approval of division manager and account manager in case of deviations;
- Organizes signing order confirmation in accordance with authorization scheme;
- Informs staff involved.

(6) Division manager
- Decides if order will be executed using project management or as routine work.

FIG. 12. Quotation qualification.

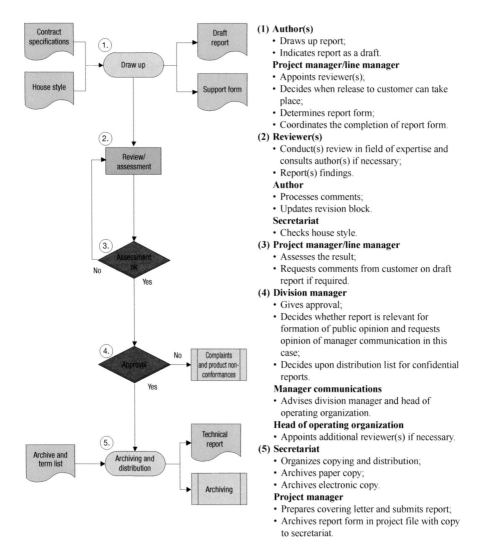

(1) Author(s)
- Draws up report;
- Indicates report as a draft.

Project manager/line manager
- Appoints reviewer(s);
- Decides when release to customer can take place;
- Determines report form;
- Coordinates the completion of report form.

(2) Reviewer(s)
- Conduct(s) review in field of expertise and consults author(s) if necessary;
- Report(s) findings.

Author
- Processes comments;
- Updates revision block.

Secretariat
- Checks house style.

(3) Project manager/line manager
- Assesses the result;
- Requests comments from customer on draft report if required.

(4) Division manager
- Gives approval;
- Decides whether report is relevant for formation of public opinion and requests opinion of manager communication in this case;
- Decides upon distribution list for confidential reports.

Manager communications
- Advises division manager and head of operating organization.

Head of operating organization
- Appoints additional reviewer(s) if necessary.

(5) Secretariat
- Organizes copying and distribution;
- Archives paper copy;
- Archives electronic copy.

Project manager
- Prepares covering letter and submits report;
- Archives report form in project file with copy to secretariat.

FIG. 13. Preparation of technical reports.

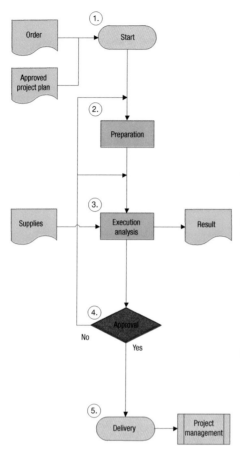

(1) **Project manager**
  - Initiates activities in accordance with project plan or order;
  - Instructs project staff.

(2) **Project team member**
  - Compiles file;
  - Selects resources to be used, such as codes, standards, etc., and defines modelling and boundary conditions, as necessary.

  Project manager
  - Determines verification path;
  - Determines extent and scope of verification;
  - Makes agreements with available verifier(s).

(3) **Project team member**
  - Compiles input;
  - Performs calculations and data processing as needed.

(4) **Verifier**
  - Verifies code selection, modelling and boundary conditions;
  - Checks input data and deliveries;
  - Verifies and discusses result with project staff and project manager.

  Project staff
  - Process verifier's proposals;
  - Report status to project manager.

  Project manager
  - Takes corrective action if necessary.

(5) **Project staff**
  - Submit results and file to project manager.

  Project manager
  - Accepts results and approves for use;
  - Archives files.

FIG. 14. Numerical calculations.

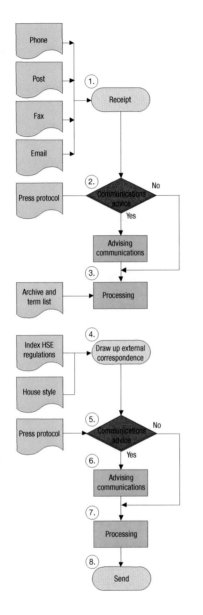

(1) Employee
- Receives and sorts mail;
- Sends original financial correspondence to financial services;
- Sends original licenses and HSE correspondence to management system representative;
- Sends project related correspondence to appropriate project manager with copy to account manager;
- Sends tender invitation to division manager;
- Sends original correspondence on personnel aspects to manager human resources.

(2) Employee
- Assesses, possibly in cooperation with line manager, whether advice of communications manager is necessary.

(3) Manager communications
- Provides advice if necessary.

(4) Responsible employee
- Arranges further handling, distribution and archiving.

(5) Employee
- Drafts a letter or fax;
- Assesses, possibly in cooperation with line manager, whether advice of communications manager is necessary.

(6) Manager communications
- Provides advice if necessary.

(7) Secretariat
- Checks house style and draws up, if necessary, letter or fax;
- Registers correspondence;
- Sends licence and HSE correspondence to management system representative for further handling;
- Sends financial correspondence to manager financial services for further handling;
- Sends correspondence on personnel aspects to manager human resources for further handling;
- Sends project related correspondence to appropriate project manager for further handling.

(8) Employee
- Coordinates signing according to authorization scheme;
- Coordinates distribution and archiving.

(9) Secretariat/employee
- Arranges sending.

FIG. 15. Incoming and outgoing correspondence.

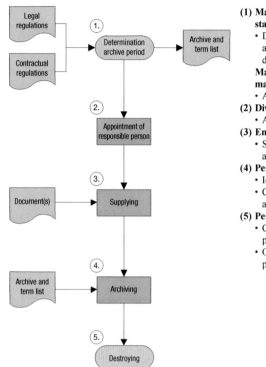

(1) **Management system representative or QSE staff member**
   • Determines and maintains administration, archive and terms list and expiration period of documents to be archived.
   Management system representative or division manager
   • Approves expiration periods.
(2) **Division manager**
   • Appoints person responsible for archiving.
(3) **Employee**
   • Supplies documents to person responsible for archiving.
(4) **Person responsible for archiving**
   • Identifies and archives documents;
   • Considers future suitability of selected archiving method.
(5) **Person responsible for archiving**
   • Checks documents at the end of the expiration period;
   • Organizes destruction after consultation with persons involved.

FIG. 16. Archiving.

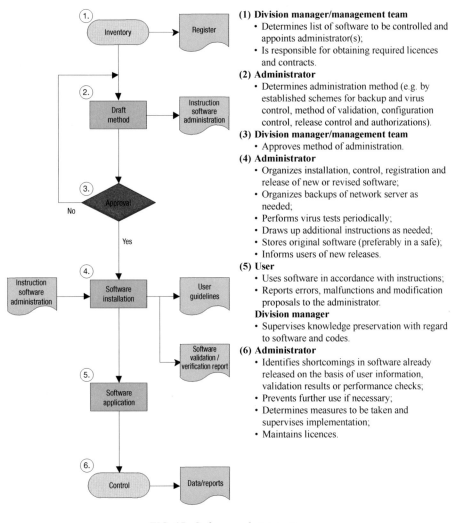

(1) **Division manager/management team**
  • Determines list of software to be controlled and appoints administrator(s);
  • Is responsible for obtaining required licences and contracts.

(2) **Administrator**
  • Determines administration method (e.g. by established schemes for backup and virus control, method of validation, configuration control, release control and authorizations).

(3) **Division manager/management team**
  • Approves method of administration.

(4) **Administrator**
  • Organizes installation, control, registration and release of new or revised software;
  • Organizes backups of network server as needed;
  • Performs virus tests periodically;
  • Draws up additional instructions as needed;
  • Stores original software (preferably in a safe);
  • Informs users of new releases.

(5) **User**
  • Uses software in accordance with instructions;
  • Reports errors, malfunctions and modification proposals to the administrator.
Division manager
  • Supervises knowledge preservation with regard to software and codes.

(6) **Administrator**
  • Identifies shortcomings in software already released on the basis of user information, validation results or performance checks;
  • Prevents further use if necessary;
  • Determines measures to be taken and supervises implementation;
  • Maintains licences.

FIG. 17. Software administration.

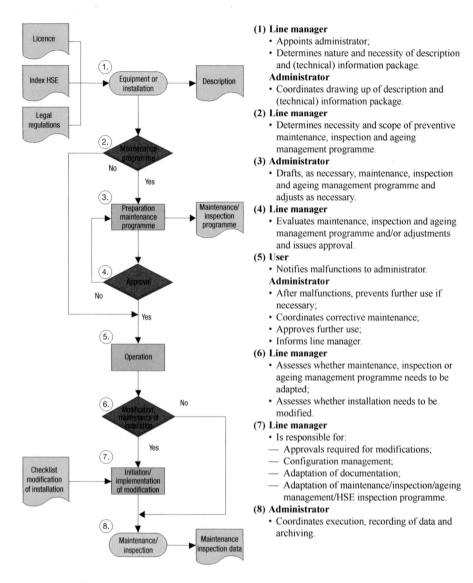

(1) Line manager
- Appoints administrator;
- Determines nature and necessity of description and (technical) information package.

Administrator
- Coordinates drawing up of description and (technical) information package.

(2) Line manager
- Determines necessity and scope of preventive maintenance, inspection and ageing management programme.

(3) Administrator
- Drafts, as necessary, maintenance, inspection and ageing management programme and adjusts as necessary.

(4) Line manager
- Evaluates maintenance, inspection and ageing management programme and/or adjustments and issues approval.

(5) User
- Notifies malfunctions to administrator.

Administrator
- After malfunctions, prevents further use if necessary;
- Coordinates corrective maintenance;
- Approves further use;
- Informs line manager.

(6) Line manager
- Assesses whether maintenance, inspection or ageing management programme needs to be adapted;
- Assesses whether installation needs to be modified.

(7) Line manager
- Is responsible for:
  — Approvals required for modifications;
  — Configuration management;
  — Adaptation of documentation;
  — Adaptation of maintenance/inspection/ageing management/HSE inspection programme.

(8) Administrator
- Coordinates execution, recording of data and archiving.

FIG. 18. Administration and maintenance of equipment and installations.

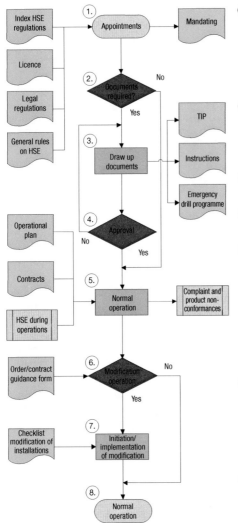

(1) Division manager
- Appoints line manager responsible for daily operations of plant/laboratories/experimental set-up/buildings;
- Establishes mandates.

Line manager
- Appoints supervisor for plants/laboratories/experimental set-up/buildings.

(2) Line manager
- Determines necessity for and nature of technical information package (TIP)/written instructions;
- Determines (emergency) drills based on emergency plan.

(3) Supervisor
- Coordinates preparation of TIP/instructions/ preventive maintenance programme.

(4) Line manager
- Approves TIP/instructions/preventive maintenance programme.

(5) Line manager
- Is responsible for operation in accordance with instructions and within technical specifications/ legal framework and licensing requirements;
- Provides additional instructions;
- Implements regular HSE inspections, tests and emergency drills;
- Is responsible for adequate handling of malfunctions;
- Initiates procedures regarding complaints and non-conformances if required.

(6) Division manager
- Evaluates operation annually with operational plan as reference;
- Initiates contract amendment if necessary, in consultation with account manager.

Line manager
- Assesses necessity of amendments in operation;
- Assesses necessity for contract amendments.

(7) Line manager
- Is responsible for:
- — Approval procedure for amendments;
- — Adaptation of documentation;
- — Adaptation of HSE inspection programme.

(8) Line manager
- Coordinates data recording;
- Supervises reporting.

FIG. 19. Operation of installations and laboratories.

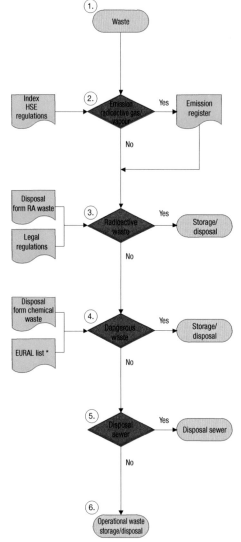

(1) Line manager
- Determines waste to be stored and/or disposed of.

(2) Line manager
- Checks if emission of radioactive gas/vapour can be expected; is responsible for registration and archiving data if required;
- Supervises storage or disposal of solids or liquids from which emission is originating.

Management system representative
- Supervises reporting of radioactive emissions.

Radiation protection supervisor
- Advises head of operating organization.

(3) Line manager/project manager
- Classifies radioactive waste in consultation with radiation protection supervisor, if necessary.

Division manager
- Organizes archiving of data with copy to radiation protection supervisor;
- Is responsible for storage and disposal in accordance with national requirements.

Radiation protection supervisor
- Records and advises head of operating organization.

Head of operating organization
- Reports to competent authorities.

(4) Line manager
- Classifies hazardous waste and is responsible for verification, storage, registration and disposal.

(5) Line manager
- Determines if discharge into sewer is allowed; if not, organizes transportation to liquid waste processing facility;
- Is responsible for registration of the flow rate and reports discharged quantities to management system representative.

Management system representative
- Supervises registration of composition and reporting of deviations.

(6) Line manager
- Classifies industrial waste and oversees disposal.

* EURAL list: European List of Waste.

FIG. 20. Waste management.

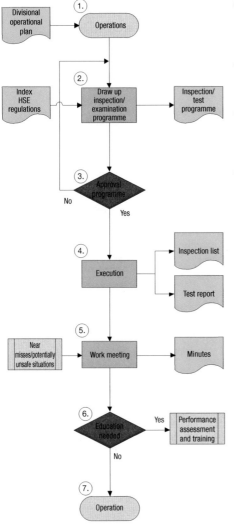

(1) Division manager
  • Determines extent of required HSE inspections and tests;
  • Appoints line manager as person responsible for HSE during day-to-day operation.
(2) Line manager
  • Draws up inspection and testing programme;
  • Appoints person responsible for inspections and tests.
(3) Division manager
  • Obtains advice from QSE representative/management system representative if necessary;
  • Approves inspection and test programme.
(4) Person responsible for inspection
  • Performs a quarterly inspection (inspection for offices annually);
  • Records non-conformances.
  Line manager
  • Performs biannual inspection programme;
  • Records deviations;
  • Analyses data from external (fire brigade) inspections;
  • Implements preventive and corrective measures;
  • Organizes filing and archiving, with copy to product group manager and QSE representative.
  Person responsible for testing
  • Performs tests;
  • Records findings.
(5) Line manager
  • Organizes the discussion of HSE subjects in progress meeting, including near misses, unsafe situations and toolbox meetings;
  • Discusses inspection/test findings and proposes measures to division manager;
  • Coordinates reporting, with copy to division manager and QSE representative.
  Division manager
  • Establishes additional measures if necessary.
(6) Line manager
  • Establishes supplementary HSE training, information and instruction in consultation with division manager and QSE representative;
  • Coordinates recording of training, information and instruction.
  Division manager
  • Implements training, information and instruction.

FIG. 21. Health, safety and environment (HSE) aspects of operations.

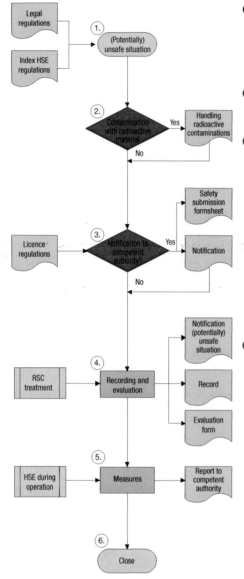

(1) Notifying employee
- Identifies a (potentially) unsafe situation or that a safety precaution or safety regulation is absent or has not functioned;
- Calls for fire brigade or first aider to assist if necessary;
- Informs line manager.

(2) Notifying employee
- Acts according to work instruction 'Handling Radioactive Contamination' in case of contamination with radioactive material.

(3) Line manager
- Assesses whether it is safe to proceed with activities;
- Assesses whether notification to the competent authorities is necessary and reports directly to division manager, head of operating organization and management system representative.

Management system representative (in case of absence: radiation protection supervisor)
- Assesses classification of notification;
- Prepares notification in cooperation with line manager.

Head of operating organization
- Sends formal notification to competent authority;
- Coordinates the external communication.

(4) Line manager
- Fills out 'Potentially Unsafe Situation' form in cooperation with notifying employee.

Management system representative
- Records and evaluates notification;
- Draws conclusions and lays down recommendations;
- Returns form to line manager, with copy to product group manager, QSE representative and notifying employee;
- Assesses whether evaluation and formal submission of evaluation report to competent authority have to take place.

Line manager
- Evaluates, if necessary, using supporting documents;
- Determines possible additional preventive measures and reports evaluation to division manager, management system representative and QSE staff member;
- Discusses evaluation during work progress meeting.

(5) Line manager
- Is responsible for implementation of measures and reports status to division manager, management system representative/QSE staff member.

(6) Management system representative
- Concludes and informs notifying employee, line manager, division manager and QSE staff member.

FIG. 22. Potentially unsafe situations.

44

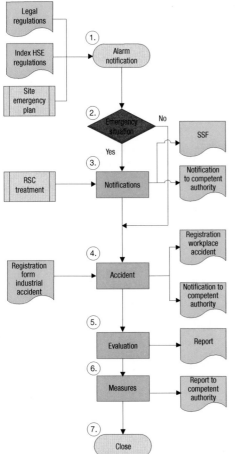

(1) Employee
- Tries to mitigate accident/emergency by local means;
- At slightest doubt calls emergency number*;
- Informs line manager.
Line manager
- Takes measures to prevent escalation and organizes guidance for emergency teams.
(2) Emergency organization
- Determines course of action and initiates it.
(3) Line manager
- Notifies head of operating organization of emergency situation immediately.
Management system representative (in absence: radiation protection supervisor)
- Prepares notification to competent authority.
Head of operating organization
- Sends notification to competent authority.
Division manager
- Organizes RSC treatment through a safety submission formsheet (SSF).
(4) Line manager
- Notifies accident by phone to head of operating organization;
- Fills out 'registration industrial accident' form in case of injury to staff, and sends it to management system representative with a copy to division manager and QSE staff member, and informs manager human resources in case of absence of employee.
Management system representative
- Records information and prepares, if necessary, notification to the competent authority.
Head of operating organization
- Sends notification to competent authority;
- Supervises external communication.
(5) Emergency services
- Report actions to head of operating organization with copy to management system representative.
Management system representative
- Assesses reports, in cooperation with involved persons, and advises the head of operating organization and division manager about possible measures to be taken;
- Advises head of operating organization with respect to formal settlement with the competent authority.
Division manager
- Determines measures in consultation with persons involved and reports to head of operating organization with copy to management system representative.
(6) Division manager
- Is responsible for implementation of measures;
- Discusses event during regular work progress meeting or begins, if necessary, a management review.
(7) Management system representative
- Informs persons involved and concludes process.

* Ensure that actual emergency number is specified in the procedure.

FIG. 23. Accident and/or emergency.

45

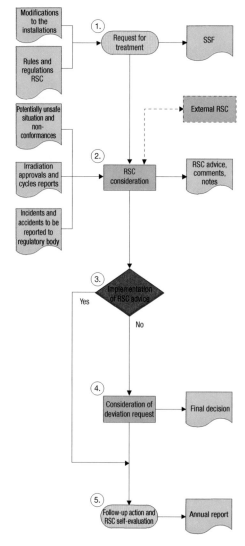

(1) Line manager/division manager or licensee
- Is responsible for preparation of documents for RSC treatment with respect to safety related aspects of operation, including:
  — New and modified experiments;
  — Modifications to the installation;
  — Non-standard safety related tests;
  — Incidents and accidents to be reported to regulatory body;
  — Safety related deviations;
  — Safety related procedures;
  — Licensing documentation.

(2) Head of operating organization
- Appoints members of external RSC to supervise and foster activities of the RSC, safety culture, and to promote implementation of internationally accepted best practices.

RSC
- Evaluates and assesses documents and reports presented;
- Requests additional information, if necessary;
- Proactively initiates safety analyses of safety related aspects as necessary;
- Involves external experts, if necessary;
- Issues written advices, comments and notes to licensees and informs the manager facility.

(3) Line manager/division manager/licensee
- Assesses RSC advice;
- Clarifies any ambiguity in cooperation with RSC;
- Implements advices and modifications.

Manager nuclear facility
- Assesses RSC advice;
- Clarifies any ambiguity with RSC;
- Implements advices and adaptations.

(4) Line Manager
- Drafts a deviation request.

Head of operating organization
- Takes final decision in case of deviation request;
- Informs RSC and line and division manager.

(5) Line manager
- Implements actions in accordance with RSC advice or with final decision.

RSC
- Reports results and self-evaluation to licensees and to external RSC annually.

FIG. 24. Reactor safety committee (RSC) evaluation.

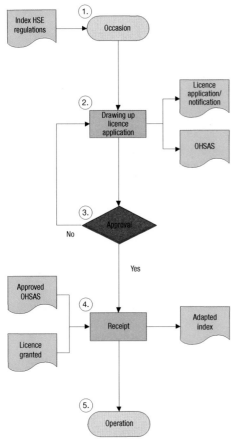

(1) **Division manager/management system representative**
  • Establishes the necessity for adaptation or application for licence and/or occupational health and safety assessment (OHSAS).
  Head of operating organization
  • Agrees with necessity for adaptation or initiation of application.

(2) **Division manager**
  • Is responsible for acquiring required licence;
  • Is responsible for the preparation of necessary supporting documents in consultation with the management system representative.
  Management system representative
  • Is responsible for the formal application and/or notification and/or OHSAS.

(3) **Division manager**
  • Is responsible for obtaining prescribed advices;
  • Approves required appendices and supporting documents.
  Reactor safety committee (RSC)
  • Advises on nuclear safety.
  Radiation protection supervisor
  • Advises on radiological aspects.
  Management system representative
  • Advises on conventional HSE issues and licencing aspects.
  Head of operating organization
  • Submits the application and/or notification and/or request for approval of OHSAS.

(4) **Management system representative**
  • Is responsible for the adaptation of licence index and/or the OHSAS index;
  • Reviews granted licence and informs division manager and head of operating organization of results;
  • Organizes archiving and distribution of licences and/or approval of OHSAS.

(5) **Division manager or head of operating organization**
  • Implements licence and/or notification and/or measures from OHSAS.

FIG. 25. Licence, occupational health and safety assessment.

6. MEASUREMENT, ASSESSMENT AND IMPROVEMENT

6.1. MONITORING AND MEASUREMENT

The Management System for Facilities and Activities [2] requires that "the effectiveness of the management system ... be monitored and measured to confirm the ability of the processes to achieve the intended results and to identify opportunities for improvement." The generic guidance that has been developed to provide a means of implementing this requirement is established in Ref. [3].

6.2. SELF-ASSESSMENT

Safety requirements established within The Management System for Facilities and Activities [2] require that "senior management and management at all other levels in the organization ... carry out self-assessment to evaluate the performance of work and the improvement of the safety culture." The generic guidance that has been developed to provide a means of implementing this requirement is established in Ref. [3]; there is no supplementary guidance.

6.3. INDEPENDENT ASSESSMENT

6.3.1. General

The Management System for Facilities and Activities [2] requires that:

"Independent assessments shall be conducted regularly on behalf of senior management:

— To evaluate the effectiveness of processes in meeting and fulfilling goals, strategies, plans and objectives;
— To determine the adequacy of work performance and leadership;
— To evaluate the organization's safety culture;
— To monitor product quality;
— To identify opportunities for improvement.

"An organizational unit shall be established with the responsibility for conducting independent assessments[11]. This unit shall have sufficient authority to discharge its responsibilities.

"Individuals conducting independent assessments shall not assess their own work.

"Senior management shall evaluate the results of the independent assessments, shall take any necessary actions, and shall record and communicate their decisions and the reasons for them.

[11] The size of the assessment unit differs from organization to organization. In some organizations, the assessment function may even be a responsibility assigned to a single individual or to an external organization."

Independent assessment includes activities such as audits or surveillances carried out to determine the extent to which the requirements for the management system are being fulfilled, to evaluate the effectiveness of the management system and to identify opportunities for improvement. They can be conducted by or on behalf of the organization itself for internal purposes, by interested parties such as customers and regulators (or by other persons on their behalf), or by independent external organizations.

The following sample process shows how internal audits may be planned and conducted in a research reactor operating organization. It is supplementary to, and is to be read in conjunction with the generic guidance established in Ref. [3].

6.3.2. Internal audits[12]

The process shown in Fig. 26 describes a methodology for drawing up an annual audit plan, appointing auditors and executing of the audit(s). It also includes the evaluation of findings, monitoring the progress of improvements and the archiving of audit reports.

6.4. MANAGEMENT SYSTEM REVIEW

Reference [2] requires that:

"A management system review shall be conducted at planned intervals to ensure the continuing suitability and effectiveness of the management system and its ability to enable the objectives set for the organization to be accomplished.

---

[12] The term 'internal audit' is used for audits performed by employees of the operating organization who do not belong to the section or division that is being audited.

"The review shall cover but shall not be limited to:
— Outputs from all forms of assessment;
— Results delivered and objectives achieved by the organization and its processes;
— Non-conformances and corrective and preventive actions;
— Lessons learned from other organizations;
— Opportunities for improvement.

"Weaknesses and obstacles shall be identified, evaluated and remedied in a timely manner.

"The review shall identify whether there is a need to make changes to or improvements in policies, goals, strategies, plans, objectives and processes."

The following sample process shows how a management system review may be planned and conducted in a research facility. It is supplementary to, and is to be read in conjunction with the generic guidance established in Ref. [3].

The process shown in Fig. 27 describes a methodology for performing a management system review at both organizational and divisional level. This review covers all aspects of the operating organization and includes, but is not limited to, improvement activities, incident reviews, accident reviews, and near miss reviews or indicators of decreasing performance of systems and persons. It includes reviews of proposed changes to the management system.

6.5. NON-CONFORMANCES AND CORRECTIVE AND PREVENTIVE ACTIONS

Reference [2] requires that:

"The causes of non-conformances shall be determined and remedial actions shall be taken to prevent their recurrence.

"Products and processes that do not conform to the specified requirements shall be identified, segregated, controlled, recorded and reported to an appropriate level of management within the organization. The impact of non-conformances shall be evaluated and non-conforming products or processes shall be either:
— Accepted;
— Reworked or corrected within a specified time period; or
— Rejected and discarded or destroyed to prevent their inadvertent use.

"Concessions granted to allow acceptance of a non-conforming product or process shall be subject to authorization. When non-conforming products or processes are reworked or corrected, they shall be subject to inspection to demonstrate their conformity with requirements or expected results.

"Corrective actions for eliminating non-conformances shall be determined and implemented. Preventive actions to eliminate the causes of potential non-conformances shall be determined and taken.

"The status and effectiveness of all corrective and preventive actions shall be monitored and reported to management at an appropriate level in the organization.

"Potential non-conformances that could detract from the organization's performance shall be identified. This shall be done: by using feedback from other organizations, both internal and external; through the use of technical advances and research; through the sharing of knowledge and experience; and through the use of techniques that identify best practices."

The following sample process shows how complaints and product non-conformances could be handled by a research reactor. It is supplementary to, and is to be read in conjunction with, the generic guidance established in Ref. [3].

6.5.1. Complaints and product non-conformances

The process shown in Fig. 28 describes a methodology for determining, handling and registering non-conformances and complaints including those from interested parties. It also includes the definition of corrective and preventive actions and independent verification of the measures taken to prevent recurrence. In addition to this process, guidance in Ref. [3] identifying determination of the cause of non-conformances and analysis of corrective actions is required to be taken into consideration.

6.5.2. Feedback from operational experience

The process shown in Fig. 29 describes a methodology for utilizing feedback from operational experience. It also includes information on how to analyse the information based on its associated risks.

6.6. IMPROVEMENT

6.6.1. General

Safety requirements established within The Management System for Facilities and Activities [2] require:

"Opportunities for the improvement of the management system shall be identified and actions to improve the processes shall be selected, planned and recorded.

"Improvement plans shall include plans for the provision of adequate resources. Actions for improvement shall be monitored through to their completion and the effectiveness of the improvement shall be checked."

The following sample process shows how improvement proposals may be handled by a research reactor. It is supplementary to, and is to be read in conjunction with, the generic guidance established in Ref. [3].

6.6.2. Improvement proposals

The process shown in Fig. 30 describes a methodology for drawing up, registering and implementing an improvement proposal. It also includes the verification of the result and the reporting of the result to the employee.

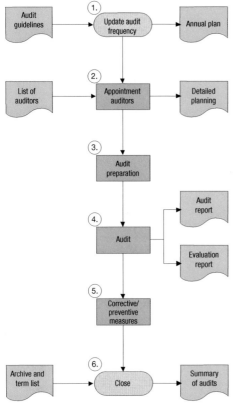

(1) Management system representative
- Draws up general guidelines for annual audit plan.

(2) Management system representative
- Draws up detailed audit plan in consultation with QSE staff members;
- Appoints auditors;
- Distributes detailed plan.

(3) Auditor
- Evaluates findings of previous audits;
- Evaluates relevant reports/documentation;
- Evaluates management review minutes;
- Evaluates complaint/non-conformance reports;
- Makes appointments with auditee.

(4) Auditor
- Performs audit; records findings, if any; requests auditees' signature for approval;
- Draws up evaluation report;
- Sends audit/evaluation report to the auditees' QSE staff member.

QSE staff member
- Records audit findings in data files and archives original audit report;
- Sends audit reports and evaluation report to auditees, line manager and to the person responsible for action;
- Sends evaluation report to management system representative.

(5) Auditee's line manager/person responsible for action
- Takes corrective or preventive actions and informs QSE staff member;
- Monitors progress of actions;
- Reports completion of actions to QSE staff member;
- Evaluates findings of previous audits;
- Evaluates improvement measures;
- Evaluates complaint/non-conformance reports.

(6) Management system representative/QSE staff member
- Is responsible for an overview of audits conducted;
- Monitors/administers audit reports;
- Verifies effectiveness of measures;
- Archives originals upon completion;
- Informs persons involved;
- Concludes procedure.

FIG. 26. Internal audits.

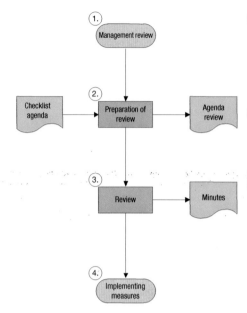

(1) **Head of operating organization/ division manager**
  • Organizes the management review; this review is clearly identified as a part of the senior management team meeting or divisional management team meeting.
  • Ensures that all relevant agenda items are discussed on an annual basis.

(2) **Management system representative/ QSE staff member**
  • Proposes agenda items for management review;
  • Evaluates last period and action points and determines trends;
  • Proposes improvements.

(3) **Head of operating organization/ division manager**
  • Assesses the effectiveness of earlier measures and decides on conclusion;
  • Determines measures for improvement of corporate management system or divisional management system, including relevant processes, product improvement according to client requirements and the need for resources;
  • Organizes the distribution of the minutes to participants, management system representative and QSE staff member.

(4) **Person responsible for action**
  • Implements measures;
  • Reports to senior management team or divisional management team and sends copy to management system representative and QSE staff member representative.

FIG. 27. Management system review.

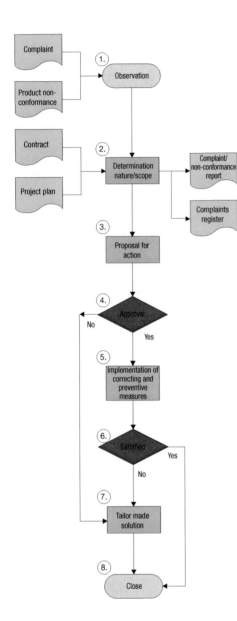

(1) Employee
- Receives complaint or observes product non-conformance;
- Notifies responsible line manager and QSE representative.

(2) Line manager
- Determines nature and scope and notifies division manager;
- Forwards confirmation within 2 weeks with deadlines and sets time limits for settlement;
- Coordinates completion of 'complaint/non-conformance' form;
- Notifies complaints from general public to management system representative and manager communications.

Manager communications
- Acts as spokesperson in case of complaints from the general public.

QSE staff member
- Registers and monitors progress.

(3) Line manager
- Informs employees involved;
- Organizes proposal to client in cooperation with QSE representative and informs division manager, account manager and QSE staff member.

(4) Line manager
- Finds out whether client accepts proposal; if client does not accept proposal, the line manager hands over the complaint to the division manager;
- Informs management system representative.

(5) Line manager
- Defines any corrective and preventive measures;
- Is responsible for implementation;
- Reports to division manager with copy to account manager and QSE staff member.

(6) Line manager
- Finds out whether client accepts the results; if client does not accept the proposal, the line manager hands over the complaint to the division manager;
- Informs management system representative;
- Informs client, account manager, management system representative and QSE staff member.

(7) Division manager
- Takes care of further conclusion, including tailor made solutions;
- Notifies completion to QSE staff member.

(8) QSE staff member
- Verifies measures taken;
- Informs persons involved and concludes procedure.

FIG. 28. Complaints and product non-conformances.

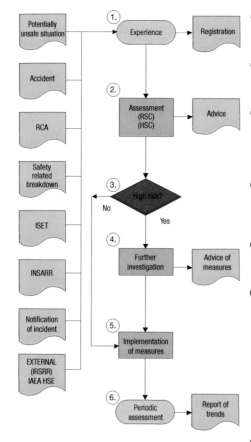

(1) Line manager
- Registers all experiences;
- Requests, if applicable, advice from RSC/ health physics safety committee.

(2) Reactor safety committee/health physics safety committee
- Assesses request and gives advice to division manager/head of operating organization.

(3) Division manager/head of operating organization
- Assesses in consultation with involved persons or experts whether there are high safety risks;
- Appoints investigation team if high safety risks are found.

(4) Investigation team
- Performs assessment and identifies improvement opportunities;
- Reports advised measures to division manager and management system representative.

(5) Division manager
- Supervises implementation of advised measures and reports this to management system representative.

(6) Management system representative
- Organizes annual evaluation and identifies trends;
- Obtains, if necessary, advice from relevant experts;
- Sends review of relevant experiences and trends, including possible improvement measures, to RSC/health physics safety committee for treatment.

Line manager
- Defines and implements improvement measures.

RCA: Root Cause Analyses
ISET: International Safety Regulation Team
HSC: High Flux Reactor (HFR) Safety Committee

FIG. 29. Feedback from operational experience.

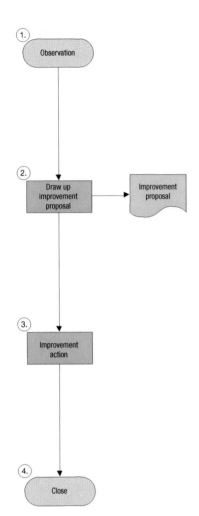

(1) Employee
- Observes that improvement is possible or necessary.

(2) Employee
- Draws up improvement proposal in consultation with person responsible for implementation;
- Consults with QSE representative, if necessary;
- Sends improvement proposal to QSE representative of person responsible for implementation;
- Indicates on improvement proposal whether it concerns a new suggestion. If so, forward proposal to the relevant committee.

QSE representative
- Records improvement proposal in database and transfers form to person responsible for indication of action and completion date.

(3) Person responsible for implementations
- Reports action and completion date and passes form to QSE representative;
- Is responsible for implementation of notified action;
- Reports completion of action to QSE representative.

QSE representative
- Records indicated action and completion date and sends copy of form to author;
- Monitors progress;
- Verifies effectiveness;
- Reports back to person responsible for implementation, if necessary;
- Records findings of effectiveness verification;
- Reports findings to author and person responsible.

(4) QSE representative
- Concludes improvement proposal;
- Informs persons involved;
- Archives relevant data.

FIG. 30. Improvement proposals.

57

7. IMPLEMENTATION OF A MANAGEMENT SYSTEM IN A RESEARCH REACTOR

7.1. GENERAL

This section provides a practical example of how an integrated management system can be implemented at an operating organization of a research reactor. Depending on its direct needs and available staff, an operating organization can select and prioritize the development and introduction of its processes using this example. Key elements for the successful implementation of an integrated management system are:

— That the person in the highest management position in the organization is responsible for ensuring that the management system is implemented.
— A collaborative effort by management, those performing the work and those assessing the work.
— The planning and deployment of adequate resources.
— That all staff are trained to achieve proficiency and to ensure they understand the procedures that apply to the performance of their work.
— Understanding that the interactions between an organization's processes can often be complex, resulting in a network of interdependent processes. Inputs and outputs of these processes are often related to both external and internal customers.
— Understanding that individual working processes rarely occur in isolation. Outputs from one process typically form part of the inputs into subsequent processes.
— Matching the complexity of the activity to the complexity of the documentation. A simple activity can be covered by one procedure while a process is implemented by the application of one or several procedures and/ or working instructions.
— Grouping several processes instead of implementing each process and related documents in a sequential manner to reduce the time and effort necessary for implementation.
— Performing dedicated internal audits to monitor and facilitate the implementation after the management system has been (partly) introduced.
— Assessing and reviewing the effectiveness of the management system at all stages of implementation.
— Using the information gained from assessments to achieve continuing improvements in work performance.

In Fig. 31, a general overview for the implementation of a management system is presented. The different items are further elaborated in the following sections. The numbers in the boxes refer to the sections where the specific step is described. The items within steps 7.2–7.4 are handled by senior management, item 7.5 by senior management and/or ad hoc groups and from 7.6 onwards by the project team.

7.2. INTRODUCTION OF THE MANAGEMENT SYSTEM

Possible inputs driving the introduction of a management system are presented in Fig. 32.

7.3. DEFINITION OF OBJECTIVES

For an operating organization, the establishment and implementation of a management system which integrates all safety, health, quality, environment, security[13] and economic elements into one coherent management system is essential in order to ensure:

— Safe operation and utilization of the research reactor facilities;
— Safe working conditions and clear working procedures;
— Compliance with licence requirements, including national and international requirements;
— Facilitation of daily work;
— Effective operation and utilization of the research reactor (scientific experiments as well as commercial activities);
— Improved communications and transparency in the organization;
— Continuous improvement.

Reference [2] requires that: "Safety shall be paramount within the management system, overriding all other demands." This can be expressed in the operating organization's policy and is one of the objectives for implementing a management system. To define the objectives of an organization, first the expectations of interested parties inside and outside the operating organization, such as customers, employees, regulatory bodies, suppliers, society and government, have to be identified. Based on the expectations of these interested parties, the objectives of the organization have to be defined by senior management.

---

[13] Usually security aspects are considered to be highly confidential and are therefore not included in the management system but rather in a confidential system applying the same general processes where applicable.

Typical objectives may be:

— To improve operational safety;
— To increase transparency;
— To improve communications and morale in the organization;
— To improve working procedures;
— To demonstrate compliance with requirements;
— To integrate safety, security, quality and environmental requirements;
— To be more efficient;
— To ensure continuous improvement;
— To reduce collective dose;
— To reduce radioactive releases.

The steps to define the objectives of an organization with the mission as a starting point is presented in Fig. 33.

7.4. MANAGEMENT COMMITMENT

Senior management have to first establish the objectives for the implementation of the integrated management system, approve the implementation plan and make the resources needed available. The objectives and the commitment of senior management have to be clearly communicated within the organization and with the key interested parties.

There must be a visible and long term senior management commitment to the establishment of a management system. Evidence of senior management commitment might be: a written statement on the foreseen implementation, active participation, periodic progress meetings, provision of the needed resources or communication to staff not directly involved. After the approval by senior management (the head of the operating organization) the decision as to which activities have to be covered by the management system, as well as management commitment to implement a process based integrated management system, will ideally to be clearly communicated to all interested parties, both internal and external. A schematic view of the involvement of senior management is given in Fig. 34.

7.5. IDENTIFICATION OF REQUIREMENTS AND STANDARDS

The Management System for Facilities and Activities [2] states that all requirements, regulations, standards and codes important for activities and for the management systems are to be identified:

"The management system shall identify and integrate with the requirements established within this publication:

— The statutory and regulatory requirements of the Member State;
— Any requirements formally agreed with interested parties (also known as stakeholders);
— All other relevant IAEA Safety Requirements publications, such as those on emergency preparedness and response and safety assessment;
— Requirements from other relevant codes and standards adopted for use by the organization."

In this section, the word 'standard' is used as a generic expression for requirements, regulations, codes, conventions, etc. An essential step in preparing the introduction of a management system is the identification and determination of the standards and requirements to be applied. The first step is to identify the applicable national, international and local legislation and agreements. All safety, quality, security, environmental and economic requirements are to be identified and, finally, the requirements of any interested parties and contractual requirements are to be identified. Senior management can also define additional requirements, which are to be incorporated in the list of requirements for the management system.

A next step for the identification of the requirements is the decision as to whether the management system is to be certified and against which requirements. Possible options are ISO 14000, OHSAS 18000 and/or ISO 9001. If certification is considered to be justified, an overview of the advantages, disadvantages and additional costs and resources has to be prepared and presented to senior management for a final decision. Finally, the proposal of the requirements to be adhered to is to be approved by senior management.

Figure 35 presents a flow chart to aid the selection of standards.

7.6. ASSIGNMENT OF THE PROJECT TEAM

The head of the operating organization has to assign a project team to plan and manage the necessary actions to develop and implement the management system. For small organizations, a project team of two or three persons is suggested; for larger organizations, this could increase to 7–9 persons.

The project team will ideally be made up of managers representing the areas of the organization involved or of representatives appointed by the managers of the different departments involved. This will give the team the knowledge and authority it needs to make proposals and devote resources to the project.

The selection of the project team members must be based on:

— Multidisciplinarity (chosen from division or line managers);
— Depth of knowledge of all processes;
— Motivation for participation.

This project team has to meet on a regular basis during the project to monitor progress, resolve questions, allocate resources and coordinate the design of the management system.

The development and implementation of a management system needs the support of all levels of the operating organization and is not only a task of the project team members — it also needs the involvement of staff members at all levels of the operating organization.

The head of the operating organization has to appoint the project leader and the team members, communicate their responsibilities and give the project leader the required authority. The project leader is typically responsible for:

— Assigning tasks for implementing changes needed in processes and procedures;
— Scheduling meetings;
— Collecting information as input for the meetings;
— Releasing meeting minutes for further distribution;
— Reporting to senior management;
— Identifying needs for the adaptation of plans.

A task group is normally set up for each process that needs to be developed and documented. The task groups examine and understand the requirements for the process and related documents. They compare the requirements of the selected standard(s) to the current process used by the organization and design a new process, or modify current ones. Once the process is approved, the task group introduce it to employees and prepare training as required. The operating organization then starts using the new process or procedure.

The process for this project team and various task groups is given in Fig. 36.

7.7. GAP ANALYSIS

The project team defines the processes and subprocesses for the organization that need to be in place to ensure that the objectives of the organization can be met and to ensure compliance with legislation, reequirements, codes and standards. The proposed processes need to be approved by senior management. After the approval of the processes, an inventory of the current status of the available procedures and instructions is performed and the gaps between the existing and the required procedures and instruction are identified. Following the gap analysis, an implementation plan has to be developed.

A flow chart to aid the performance of the gap analysis is given in Fig. 37.

7.8. STRATEGY FOR IMPLEMENTATION

The strategy put forward in this example is to develop the management system on a step-by-step basis introducing each process as it is developed and when it has the necessary support infrastructure to be implemented.

Elements of this strategy are:

— Integrate in existing system (the existing system usually already contains all operating procedures);
— All employees must be involved;
— Keep it simple and practical;
— Group related processes;
— Set priorities;
— Prepare a schedule;
— Adhere to the schedule;
— Seek improvements while implementing.

It is recommended to implement the management system in this way and not to wait for the full system to be developed and then to implement all the processes at the same time. Implementation will ideally begin as soon as possible once each individual process is completed.

When implementing a management system, it is often advisable to consider the 'Plan–Do–Check–Act' cycle, as shown in Fig. 38, which can be used to develop and implement each of the organization's processes.

Plan — Establish the objectives and processes necessary to deliver results in accordance with customer requirements and the organization's policies.

Do — Implement the processes.

Check — Monitor and measure processes and product against policies, objectives and requirements for the product and report the results.

Act — Define and implement actions to continually improve process performance.

7.9. RESOURCES

The project team have to identify the resources needed for implementing the management system such as:

— Human resources (including consultants);
— Person-hours;
— Training (internal/external);
— Financial resources.

The establishment of a management system is a time intensive job and it is important not to underestimate the resources needed.

The head of the operating organization has to ensure the required resources are provided.

7.10. SCHEDULE AND MILESTONES

In order to ensure that implementation can proceed, a time based schedule has to be established by the project team, taking into account the availability of resources and the ongoing commitments and activities of the organization. This time schedule has to include as a minimum:

— Dates (e.g. start, meetings, reporting, distribution);
— Deadlines;
— Preparation;
— Review;
— Authorization;
— Milestones.

The use of a four phased approach to develop each process is often beneficial for implementation:

— Phase 1: Introduction;
— Phase 2: Definition of activities;
— Phase 3: Implementation and communication;
— Phase 4: Follow up and adjust.

In establishing the phases of different processes, the interdependency and interrelationship between the processes has to be taken into consideration.

Typical examples of meeting schedules are:

— Project team meeting every month;
— Task group meetings every 14 days.

A sample schedule for the implementation of various processes is given in Fig. 39.

7.11. PHASES IN DEVELOPMENT

The phases in the development steps for each process are schematically presented in Fig. 40. There have to be measurable and quantifiable improvement targets.

7.12. MANAGEMENT SYSTEM MANUAL

The top tier document in a management system that provides the overview of how the organization and its management system are designed to meet its policies and objectives can have many different titles, such as 'Nuclear Management System Manual', 'Management System Description', 'QA Programme', 'QA Manual' and 'Integrated Management System Manual'. It will ideally be clear from the hierarchy of information in the organization what the name of the top tier document is. The following example refers to a management system manual.

The phases in the development of a manual are schematically presented in Fig. 41.

7.13. IMPROVEMENT OF THE MANAGEMENT SYSTEM

Continual improvement of the processes of an organization can lead to enhanced safety performance and efficiency benefits such as cost reductions and improved cycle times.

In Section 6 of this publication, the most important processes for the improvement of the management system are presented. Besides these processes, other self-assessment aspects such as facility walkdown, work observation and staff meetings can provide useful input to improve the management system.

Text cont. on p. 74.

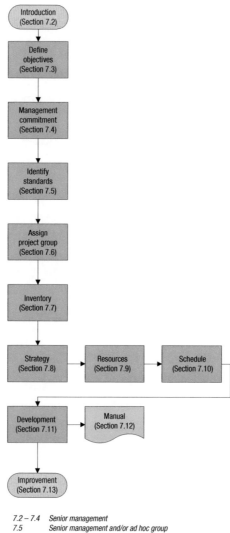

7.2 – 7.4   Senior management
7.5   Senior management and/or ad hoc group
7.6 – 7.13 Project group

FIG. 31. Implementation flow chart for a management system.

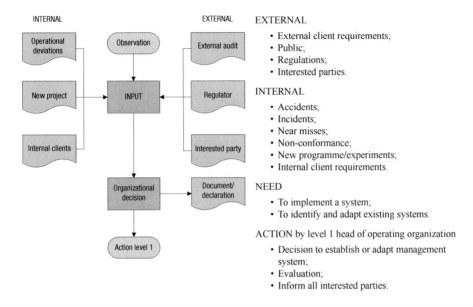

INTERNAL   EXTERNAL

EXTERNAL

- External client requirements;
- Public;
- Regulations;
- Interested parties.

INTERNAL

- Accidents;
- Incidents;
- Near misses;
- Non-conformance;
- New programme/experiments;
- Internal client requirements.

NEED

- To implement a system;
- To identify and adapt existing systems.

ACTION by level 1 head of operating organization

- Decision to establish or adapt management system;
- Evaluation;
- Inform all interested parties.

FIG. 32. Inputs driving the introduction of a management system.

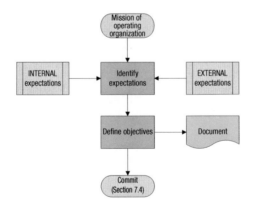

FIG. 33. Objectives for the implementation of a management system.

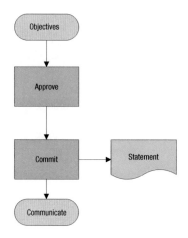

FIG. 34. Involvement of senior management.

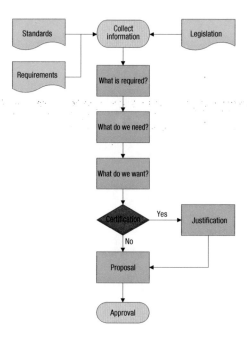

FIG. 35. Selection of requirements, codes and standards.

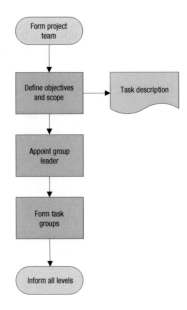

FIG. 36. Assignment of the project team.

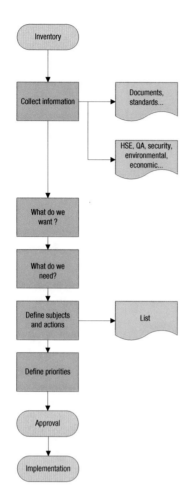

The project team is to consider the following:

- Practical and simple approach;
- What do we need?
- What do we have?
- What is available and what is missing (e.g. documents, processes, provisions, resources, etc.)?

The project team may use the following inputs:

- Self-assessment;
- Assessment by an external organization;
- Customer related processes;
- Design and/or development;
- Production and service operations;
- Control of measuring and monitoring.

The gap analysis will result in a number of management system procedures to be developed.

The action list is approved by the head of the operating organization.

FIG. 37. Gap analysis of available management system processes.

70

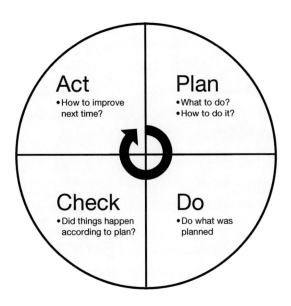

FIG. 38. Plan–Do–Check–Act cycle.

Month	Process 1	Process 2	Process 3	Process 4	Process 5	Process 6	Process *n*
0–2	Phase 1						
2–4	Phase 2	Phase 1					
4–6	Phase 3	Phase 2	Phase 1				
6–8	Phase 4	Phase 3	Phase 2	Phase 1			
8–10		Phase 4	Phase 3	Phase 2	Phase 1		
10–12			Phase 4	Phase 3	Phase 2	Phase 1	
12–14				Phase 4	Phase 3	Phase 2	Phase 1

FIG. 39. Schedule for implementation.

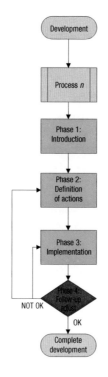

Phase (1) Introduction of a process or procedure (familiarization, collect, analyse):
- Documentation of the process or procedure;
- Define standards for existing and new documents;
- Read documentation;
- Interpretation.

Phase (2) Definition of activities:

- Timing;
- Tasks;
- Practical items;
- Define procedure;
- Authorization;
- Milestones and schedule.

Phase (3) Implementation/communication: implement the new management system procedure through communication by the group leader and involved managers to everybody:

- Posters;
- Meetings;
- Distribution;
- Training.

Phase (4) Follow-up and adjust: to promote the follow-up and adjustment; the involvement of all project team members is needed. The group leader is responsible for collecting all data.

Follow-up by:
- Suggestions/response;
- Feedback from users;
- Practical performance;
- Review;
- Etc.

FIG. 40. Phases in development.

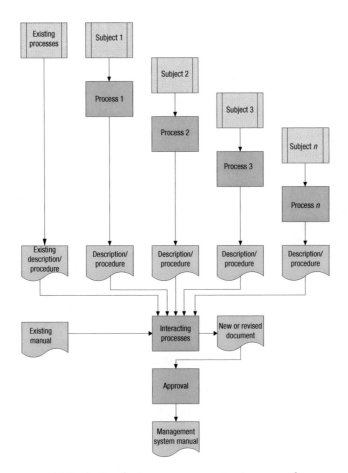

FIG. 41. Developing a management system manual.

8. SUMMARY AND CONCLUSIONS

The global research reactor community includes a highly diverse number of facility designs, organizational structures and technical missions. Research reactor power levels range from zero power critical facilities to over 100 MW. Some research reactor operating organizations are large and are comprised of teams of operators, maintenance technicians, safety and radiation control officers, managers, licensing and other support staff. Other reactors are successfully operated by a relatively small team of around ten permanent staff or even fewer. Similarly, technical duties can be very diverse, complex or rather simple, including a limited number of very specialized applications.

Management systems for larger research reactor organizations will have similarities with systems developed for nuclear power plants with comparable organizational structures. Significant differences can arise from the more diverse customer base managed by an organization operating a large multipurpose research reactor. Here, customers may range from technical industries, educational institutions, national and international research organizations, water and agricultural management programmes and even national nuclear energy programmes. Smaller research reactor organizations often also have a diverse community of customers, but execute a more straightforward utilization programme managed by a much smaller team.

In the case featured in Annex I, an integrated management system developed for a smaller research reactor operating organization applies a graded approach to ensure that all requirements are fully satisfied, but with significantly fewer staff. In this case, responsibilities are distributed among available staff with each person filling the role of several different positions and duties. The system is established to ensure independence, high standards of safety, quality and reliability, and continuous improvement. Since many research reactors are managed by small organizations, finding an example of the development and implementation of a management system for this case was key to producing comprehensive and effective guidance for the research reactor community.

The specific experience described in Annex I constitutes an example of management system implementation conforming to both legal and user requirements in a typical medium sized research reactor.

Among others, aspects typically related to such organizations have been considered, e.g.:

— Limited staff number;
— Multiple responsibilities/duties for the same personnel unit;
— Need for multidisciplinary skills;

— Capability to develop an adequate graded approach in each specific situation;
— Capability to perform process measurements and assessment;
— Limited financial resources.

In conclusion, the choice of implementing an integrated management system able to satisfy both national and international compulsory requirements in conformance with IAEA Safety Standards Series No. GS-R-3 (i.e. safe reactor operation and maintenance) and typical ISO 9001 requirements (e.g. continuous improvement, the satisfaction of interested parties and users, etc.) is discussed.

Through this systematic and graded approach, which led to the standardization of the processes involved in reactor operation and maintenance, the aspects of reactor management mentioned in Ref. [1] were addressed. This publication provides guidance for establishing, implementing, assessing and continuously improving a management system for facilities and activities that integrates safety, health, environmental, security, quality and economic elements.

DEFINITIONS

The following definitions are taken from Ref. [14].

facilities and activities. A general term encompassing nuclear facilities, uses of all sources of ionizing radiation, all radioactive waste management activities, transport of radioactive material and any other practice or circumstances in which people may be exposed to radiation from naturally occurring or artificially radioactive sources.

independent assessment. Assessments such as audits or surveillances carried out to determine the extent to which the requirements for the management system are fulfilled, to evaluate the effectiveness of the management system and to identify opportunities for improvement. They can be conducted by or on behalf of the organization itself for internal purposes, by interested parties such as customers and regulators (or by other persons on their behalf), or by external independent organizations.

management system. A set of interrelated or interacting elements (system) for establishing policies and objectives and enabling the objectives to be achieved in an efficient and effective manner. The management system integrates all elements of an organization into one coherent system. These elements include the organizational structure, resources and processes. Personnel, equipment and organizational culture as well as the documented policies and processes are parts of the management system. The organization's processes have to address the totality of the requirements on the organization as established in, for example, IAEA Safety Standards and other international codes and standards.

management system review. A regular and systematic evaluation by senior management of an organization of the suitability, adequacy, effectiveness and efficiency of its management system in executing the policies and achieving the goals and objectives of the organization.

regulatory body. An authority or a system of authorities designated by the government of a State as having legal authority for conducting the regulatory process, including issuing authorizations, and thereby regulating nuclear, radiation, radioactive waste and transport safety. The national competent authority for the regulation of radioactive material transport safety is included in this description, as is the Regulatory Authority for radiation protection and safety.

(nuclear) safety. The achievement of proper operating conditions, prevention of accidents or mitigation of accident consequences, resulting in protection of workers, the public and the environment from undue radiation hazards.

safety culture. The assembly of characteristics and attitudes in organizations and individuals which establishes that, as an overriding priority, protection and safety issues receive the attention warranted by their significance.

self-assessment. A routine and continuing process conducted by senior management and management at other levels to evaluate the effectiveness of performance in all areas of their responsibility.

Appendix

ABBREVIATIONS FOR SAMPLE PROCESSES

The following abbreviations have been used in the example processes provided:

DM	Division manager
FS	Financial services
HR	Human resources
HSE	Health, safety and environment
NPP	Nuclear power plant
OHSA	Occupational health safety assessment
QSE	Quality, safety and environment

LEGEND FOR SAMPLE PROCESSES

General

— Every document is monitored through an index. The index concerned indicates the version currently valid.
— A new revision of the index is issued when a new or revised document is issued. Modifications are indicated on the index.
— Authorization of a document takes place through the index.
— The recipient of a new revision is responsible for destroying the expired index and the expired documents.
— Input documents appear on the left and output documents at the right hand side.
— In a procedure, each document appears only once as an input document and only once as an output document, even when it is more frequently used as either an input or output, respectively.
— Decisions, releases or authorizations need to be recognisable and performed tangibly. This could be by signing a form or in another way, such as, for example, through a meeting report.
— In the procedures, general names are used for functions. If a different name is used within a division, then this has to be defined through a cross-reference table, which is valid for that division.
— Mandates of responsibilities need to be approved by the head of the operating organization and must be recorded. The mandating official retains responsibility for the correct completion of his or her share in the procedure.
— Delegation of tasks and responsibilities needs to be recorded. In this case, the delegating official also remains responsible for the correct completion of his or her share in the procedure.
— By 'division' or 'division manager', 'staff group' or 'staff group manager' is intended where applicable.

Flow diagram

First or last activity of a procedure

Action or activity

Decision

Document

Other procedure

Undefined information source.

Suggestions for improvement

The documents are adapted regularly. All staff are invited to provide feedback or to submit an improvement proposal. Such contributions will help to keep the system up-to-date and will help to implement further improvement.

Remark: Some standards such as ISO 9001 require a description of how to deal with objects supplied by customers. A minimal description to fulfil this requirement could be:

'If a customer provides us with objects in order to fulfil his or her order, then the objects will be registered at reception. Their storage, location and consumption, if any, will be recorded. The owner will be informed of any issues relating to the object. This is valid for samples and test objects to be returned.'

REFERENCES

[1] INTERNATIONAL ATOMIC ENERGY AGENCY, Safety of Research Reactors, IAEA Safety Standards Series No. NS-R-4, IAEA, Vienna (2005).

[2] INTERNATIONAL ATOMIC ENERGY AGENCY, The Management System for Facilities and Activities, IAEA Safety Standards Series No. GS-R-3, IAEA, Vienna (2006).

[3] INTERNATIONAL ATOMIC ENERGY AGENCY, Application of the Management System for Facilities and Activities, IAEA Safety Standards Series No. GS-G-3.1, IAEA, Vienna (2006).

[4] INTERNATIONAL ATOMIC ENERGY AGENCY, The Management System for Nuclear Installations, IAEA Safety Standards Series No. GS-G-3.5, IAEA, Vienna (2009).

[5] INTERNATIONAL ATOMIC ENERGY AGENCY, Use of a Graded Approach in the Application of the Safety Requirements for Research Reactors, IAEA Safety Standards Series No. SSG-22, IAEA, Vienna (2012).

[6] INTERNATIONAL ATOMIC ENERGY AGENCY, The Operating Organization and the Recruitment, Training and Qualification of Personnel for Research Reactors, IAEA Safety Standards Series No. NS-G-4.5, IAEA, Vienna (2008).

[7] INTERNATIONAL ATOMIC ENERGY AGENCY, Safety in the Utilization and Modification of Research Reactors, IAEA Safety Standards Series No. SSG-24, IAEA, Vienna (2012).

[8] INTERNATIONAL ATOMIC ENERGY AGENCY, Managing Change in Nuclear Utilities, IAEA-TECDOC-1226, IAEA, Vienna (2001).

[9] INTERNATIONAL ATOMIC ENERGY AGENCY, Configuration Management in Nuclear Power Plants, IAEA-TECDOC-1335, IAEA, Vienna (2003).

[10] INTERNATIONAL ORGANIZATION FOR STANDARDIZATION, Quality Management Systems — Guidelines for Configuration Management, ISO 10007:2003 ISO, Geneva (2003).

[11] INTERNATIONAL ATOMIC ENERGY AGENCY, Ageing Management for Research Reactors, , IAEA Safety Standards Series No. SSG-10, IAEA, Vienna (2010).

[12] INTERNATIONAL ATOMIC ENERGY AGENCY, Radiation Protection and Radioactive Waste Management in the Design and Operation of Research Reactors, IAEA Safety Standards Series No. NS-G-4.6, IAEA, Vienna (2009).

[13] INTERNATIONAL ATOMIC ENERGY AGENCY, Operational Limits and Conditions and Operating Procedures for Research Reactors, IAEA Safety Standards Series No. NS-G-4.4, IAEA, Vienna (2008).

[14] INTERNATIONAL ATOMIC ENERGY AGENCY, IAEA Safety Glossary: Terminology Used in Nuclear Safety and Radiation Protection: 2007 Edition, IAEA, Vienna (2007).

Annex I

A CASE STUDY: IMPLEMENTATION OF
A MANAGEMENT SYSTEM FOR
THE TRIGA MARK II RESEARCH REACTOR
AT THE LABORATORY OF APPLIED NUCLEAR ENERGY (LENA)
OF THE UNIVERSITY OF PAVIA, ITALY

I–1. INTRODUCTION

I–1.1. General

This annex provides an example for the implementation of a management system for operating organizations of research reactors, based on a case study in which the implementation of such a system has been completed.

The case study relates the experience of the Applied Nuclear Energy Laboratory (hereafter referred to as LENA) of the University of Pavia, Italy.

This example is used because of the recent completion of the implementation of an integrated management system, and also because of the specific characteristics of the organization (such as the limited number of staff, limited financial resources, etc.), which are often typical for organizations that operate smaller research reactors.

Section I–1 gives a brief presentation of the organization, including the scope of work, the main activities performed, the organizational structure, the identification of interested parties and the applicable requirements and standards. Section I–2 describes the LENA Management System, the reasons for its implementation, the stages of its development and the processes involved.

Some practical examples related to the development of the LENA Management System are discussed in Section I–3, indicating the choices made by the organization. In particular, Section I–3.12 shows the correlation between the LENA Management System processes and the processes considered in the main body of this publication.

I–1.2. LENA profile

LENA is an interdepartmental research centre of the University of Pavia, which operates, among other facilities, a 250 kW TRIGA Mark II research reactor. A view of the reactor hall can be seen in Fig. I–1 and the reactor core is shown in Fig. I–2.

The reactor is at the disposal of researchers from the University of Pavia and other users, both public and private, for research activities, training and education and other services. The centre itself carries out research and training activities and provides services to private enterprises, encouraging the transfer of the results of nuclear technology research to production systems, including the education and training of specialists in nuclear technology.

In particular, owing to the expertise gained in the management of the reactor and the possibility of using a research reactor and a hot laboratory for the handling of radioisotopes, LENA offers training activities developed and structured for different target groups (e.g. large corporations, universities and governmental institutions) with the objective of providing education at different learning levels characterized by a significant experimental component.

The reactor reached its first criticality on 15 November 1965. In over 45 years of operation, more than 600 publications have been issued in relation to the various fields of research conducted with the reactor, such as nuclear chemistry and radiochemistry, activation analysis, chemistry, nuclear physics, reactor physics and dosimetry.

More than forty-five years after its first criticality, the reactor is in excellent condition and is still used for important research, including:

— Experimental cancer therapies for the treatment of multifocal liver tumours by irradiation with neutrons, known as boron neutron capture therapy (BNCT);
— Analysis of materials and environmental samples for the determination of trace elements by neutron activation analysis (NAA);
— Analysis of food samples by NAA for the purpose of food safety and quality;
— Study of radiation induced damage to electronic components for aerospace applications and in particle accelerators;
— Production of radioisotopes and activation of components for industrial non-destructive diagnostics;
— Production of branded products for medical diagnosis;
— Dating of materials and products and determination of their geographical origin;
— Study of the characteristics of materials subjected to radiation and the changes in their lattice structures;
— Induction of structural changes in new materials (especially semiconductors and superconductors) and their characterization;
— Studies to characterize materials with respect to their nuclear properties.

I–1.3. Organizational structure

The LENA organizational structure, approved by the Italian regulatory body, is defined in accordance with the statutory and regulatory requirements for reactor operation, as well as for nuclear safety and health protection purposes. LENA integrated this organizational structure with the inclusion of all functions involved in the management system; the result is a comprehensive organizational chart containing both functions (see Fig. I–3) and names. A short description of the main job functions shown in the organizational chart is given below.

Director. The LENA Director is responsible for the technical and administrative management of the TRIGA Mark II reactor and the annexed radiochemical laboratory and is responsible for personnel management. The Director is also delegated by the holder of the reactor operating licence (the Rector of the University) regarding the responsibility and specific obligations of the plant in the field of nuclear safety and health protection. As required by national legislation, the Director holds a licence for nuclear plant technical management (technical director), issued by the Italian Ministry of Labour and Social Policy. The Director is also the chief technical director of the reactor.

To ensure coverage at all times by the necessary technical and organizational management, the Director establishes a call for duty shifts among staff in possession of a licence for nuclear plant technical management (technical director), issued by the Italian Ministry of Labour and Social Policy (as required by national legislation).

Deputy Director. The Deputy Director works with the Director to manage the reactor. In the absence of the Director, the management of LENA is assumed by the Deputy Director.

Technical directors. All personnel holding a licence for nuclear plant technical management (technical director), issued by the Italian Ministry of Labour and Social Policy. As defined by law, a technical director is responsible for the completion of the technical aspects relating to the determination, organization and coordination of activities related to the operation of the plant as well as the execution and the supervision of the controls concerning nuclear safety and health protection.

Reactor supervisors. Personnel holding a first level licence for technical management of a nuclear facility (technical director) issued by the Italian Ministry of Labour and Social Policy. In case of the absence of the Director and Deputy Director, the technical director on shift is responsible for the technical management of the nuclear facility, but this responsibility does not extend to administrative aspects.

Reactor operators. Personnel in possession of a second level licence for reactor operation issued by the Italian Ministry of Labour and Social Policy.

Operators are employed to operate and control the reactor during operations including handling of fuel, handing and operation of experimental set-ups, performing surveillance testing within his or her competence, informing the supervisor on duty of any anomalies, etc.

Radiation Protection Advisor. The Radiation Protection Advisor has the responsibility of monitoring the physical protection of workers and population around the facility on behalf of the employer within the powers conferred by law.

Health physicist. The Health Physics Section carries out enforcement activities related to the surveillance of the physical protection of workers and the general public, as well as environmental monitoring.

Mechanical maintenance. The Mechanical Maintenance Section carries out all activities related to ordinary and extraordinary maintenance of the equipment under its responsibility. The Mechanical Maintenance Section is entrusted with the surveillance, maintenance and efficiency of the following:

— Fire detection and extinguishing system;
— Air filtration and air conditioning system;
— Emergency air exhaust system;
— Primary and secondary cooling system;
— Demineralized water production and storage;
— Bridge crane;
— Other mechanical equipment.

Instrumentation and Electrical Maintenance. The Instrumentation and Electrical Maintenance Section carries out preventive and corrective maintenance on the following equipment:

— Electrical system;
— Area monitoring systems (ionizing radiation);
— Reactor control and monitoring system;
— Radiation protection instrumentation;
— Security systems (intrusion and access control systems, CCTV systems);
— Work equipment, with regard to electrical hazards.

The Instrumentation and Electrical Maintenance Section is responsible for the surveillance programme of the Instrumentation and Electrical installation, including calibration, verification, adjustment and repair of the measuring instruments, software, standards and reference materials necessary to ensure reliable information. The section is also responsible for the scientific instrumentation used at the reactor for performing experiments.

Quality Assurance Unit. The Quality Assurance Unit provides the necessary support for the management of the organization, with the aim of identifying and implementing strategies, processes and measures necessary to achieve LENA QMS goals and, in particular, those related to Nuclear Safety and Health Protection.

Authorized Chemist. The Authorized Chemist advises on the compatibility of materials to be irradiated in the reactor.

Communication and Marketing. The Communication and Marketing Office is dedicated to communication with interested parties and customers, and to the promotion of reactor activities.

Analysis and Measurement. The Analysis and Measurement Service performs radiochemical and spectrometric analysis for the research reactor and for external customers.

As is typical for smaller research reactors, the number of personnel is limited and therefore the various tasks and functions in the organization must be combined. Currently, the staff on duty at the LENA reactor consists of 16 employees.

Each section or unit of the organizational chart is usually made up of a head of section or unit and one or more employees. This enables every staff member to be involved in more than one task, namely the allocation of multiple responsibilities to the same person (while avoiding conflicts of interest). For example, at LENA one of the reactor managers is also a reactor supervisor and Head of the Quality Assurance Unit. The allocation of LENA personnel according to the different functions is reported in Tables I–1 and I–2.

Table I–2 underlines the multiple responsibilities assigned and shows that almost every staff member is involved in more than one function. The total number of technical staff at LENA (15) is significantly lower than the total number of the different reactor functions (34).

This type of organization, which stems from the need to better manage the limited human resources available, has the positive effect of increasing the skills of staff in the various plant activities, allowing for shift turnover and ensuring the reliability of services related to nuclear safety, health protection and the operation of the reactor itself at all times.

Together with the limited human resources, another critical point for research reactors may be the availability of the financial resources necessary to ensure compliance with all the safety requirements for reactor operation. In this regard, one of the strategies adopted by LENA is to raise income from the expertise and the continuous training of the staff.

Indeed, this internal competence enables the operating organization to become more independent in terms of operating costs, for example, with regard to regular and corrective maintenance, radiological evaluations and the development of the reactor management system.

TABLE I–1. NUMBER OF STAFF IN POSSESSION OF A LICENCE FOR REACTOR TECHNICAL MANAGEMENT OR FOR REACTOR OPERATION

Licensed reactor operation and management staff	
Reactor manager	3
Reactor supervisors	6
Reactor operators	6

TABLE I–2. ALLOCATION OF LENA STAFF TO DIFFERENT DUTIES

Functions		No. of staff involved
Administrative staff		1
Technical staff		15
of which:	Reactor manager	3
	Reactor operation and management	12
	QA	2
	Radiation Protection Advisor	1
	Health Physics Section	7
	Mechanical maintenance	2
	Electrical systems and instrumentation maintenance	2
	Analysis	3
	Marketing and customer	1
Total No. of staff		16
Total number of allocations to reactor functions		34

In order to better explain how the typical management system functions for a research reactor are allocated at LENA, Table I–3 shows the correlation between functions described in the main body of this publication (see Sections 2–6) and the corresponding LENA functions.

TABLE I–3. TYPICAL MANAGEMENT SYSTEM FUNCTIONS FOR A RESEARCH REACTOR

LENA function	LENA QMS function as described in the main body (see Sections 2–6)
Director/Reactor manager	Division manager Reactor manager Manager human resources
Head of Quality Assurance	Management system representative Manager QSE Person responsible for archiving Auditor
Radioprotection Advisor	Radiation protection supervisor
Reactor Supervisor	Person responsible for inspection
Reactor Operator	Responsibilities normally described in operator instructions and not part of the procedures in the main document
Head of Mechanical Maintenance	Line/task manager Person responsible for inspection Purchaser
Head of Instrumentation and Electrical Maintenance	Line/task manager Person responsible for inspection Purchaser
Head of Analysis and Measurements	Line/task manager Purchaser
Head of Health Physics	Line/task manager Person responsible for inspection Purchaser
Communication and Marketing	Manager communications
Administration	Secretariat Manager financial services

I–1.4. Identification and analysis of interested parties

In relation to the management and operation of a nuclear facility, interested parties are usually defined as those who may have a specific interest in a given issue or decision. They may be either internal (those involved in the management and operational processes) or external (those who may be affected by the potential outcome of facility management).

In order to develop the management system, the first task was to identify the interested parties. In fact, only by establishing exactly who the interested parties are, is it possible to identify all the requirements that must be met by the operating organization and to plan the necessary actions to ensure the management of the processes involved. They can be grouped as follows:

— **Primary interested parties** are those ultimately affected, either positively or negatively, by an organization's actions; they can be both internal or external to the organization. In this specific case, these are:
  • Workers;
  • Employer/owner;
  • Users/customers.
— **Secondary interested parties** are the 'intermediaries', that is, persons or organizations who are indirectly affected by an organization's actions (external to organization):
  • Civil society;
  • General public.
— **Key interested parties** have significant influence upon or importance within an organization:
  • Regulatory body;
  • National organizations/ministries;
  • International organizations.

Each interested party has several requirements, which may be legal requirements (as in the case of the regulatory body) or not (as in the case of users/customers). The relationship between the various interested parties and the most relevant requirements is shown in Fig. I–4.

In summary, the satisfaction of the requirements of interested parties is achieved by ensuring the fulfilment of the objectives below throughout the management and operation of the reactor.

— Safety (in the definition used at LENA, safety includes nuclear safety, occupational health and protection of the people, workers and environment);

— Security (Security is included in the Integrated Management System, but it is excluded from the scope of certification in relation to management of confidential data and information);
— Efficiency;
— Effectiveness.

For a more detailed explanation about the strategy that the organization decided to follow in order to achieve these objectives, see Section I–2.1.

I–1.5. Applicable requirements and standards

An organization that manages a reactor has to deal with a large number of laws, regulations, codes, conventions, etc.; they are identified here by the word 'standard'. The identification of standards allows the organization to identify the applicable requirements, which are essential in setting up a management system.

For this reason, LENA has compiled a list of all standards grouped by categories; Table I–4 provides some examples of the specific standards with which the organization has to conform.

I–2. LENA MANAGEMENT SYSTEM

I–2.1. Scope and application field of LENA management system

As previously mentioned, the research reactor is used for many different purposes (basic research, applied research and technology transfer, education and training and public information), and addresses various types of users (schoolchildren, university students, researchers, companies, workers, individuals).

Furthermore, in Italy, a research reactor is considered by law identical to a nuclear power plant in all aspects, so it must adhere to all applicable laws and regulations in the field of nuclear safety and protection of the environment, population and workers.

Based on the above, the reactor organization needs to manage its processes both to meet the demands arising from specific purposes and users, and to ensure the highest mandatory safety standards: this is the aim and scope of the management system as it is implemented at the LENA reactor.

At the LENA reactor the management system is referred to as the quality management system (QMS). It aims to be the equivalent of what is usually called an integrated management system, which takes into account all safety, quality, security and environmental issues.

TABLE I–4. MAIN APPLICABLE REQUIREMENTS AND STANDARDS

Standard category	Examples (not exhaustive)
International directives and regulations	• Code of Conduct on the Safety of Research Reactors [I–1]; • Treaty on the Non-proliferation of Nuclear Weapons [I–2]; • Convention on the Physical Protection of Nuclear Material [I–3].
Statutory and regulatory National (Italian) standards	• Framework Act on the Peaceful Use of Nuclear Energy (No. 1860, 1962-12-31); • Regulation for the Recognition of the Fitness For Purpose of the Technical Operation of Nuclear Installations (No. 1450, 1970-12-30); • Implementation of the Council of the European Union Directive laying down basic safety standards for the protection of the health of workers and the general public against the dangers arising from ionizing radiation (96/29/EURATOM, 1996-05-13; Italian Legislative Decree 241, 2000-05-26).
Specific standards for the research reactor organization (issued by Italian regulatory body or ministries)	• Operation Licence of the TRIGA Mark II reactor • TRIGA Mark II reactor Operating Rules • TRIGA Mark II reactor Management Rules • TRIGA Mark II reactor Operational requirements
Not legal requirements, but to be used as guidelines	• IAEA Safety Standards for Research Reactors, such as: • SAFETY REQUIREMENTS (e.g. Safety of Research Reactors, NS-R-4 [I–4]; The Management System for Facilities and Activities GS-R-3 [I–5]) • SAFETY GUIDES (e.g. Application of the Management System for Facilities and Activities, GS-G-3.1 [I–6]; The Management System for Nuclear Installations, GS-G-3.5 [I–7]; Maintenance, Periodic Testing and Inspection of Research Reactors, NS-G-4.2 [I–8], etc.) • TECHNICAL REPORTS (e.g. Optimization of Research Reactor Availability and Reliability: recommended practices, NP-T-5.4 [I–9])
Standards defined by the organization	• Internal procedures and instructions • ISO Standards (ISO 9000 [I–10], ISO 9001 [I–11], ISO 17025 [I–12], ISO 10012 [I–13], etc.) • Internal University regulations

In fact, the organization unequivocally puts nuclear safety first, prioritizing it over all other needs. The LENA QMS is intended as a tool to enable the continued development of safety culture and to achieve higher levels of security.

All aspects of nuclear safety, health protection and environmental protection are included in LENA's QMS and safety culture is present and integrated in all activities, e.g.:

— *Definition of responsibilities:* Both the Director and staff at all levels are aware of their responsibilities and that their actions affect the safety of the activities connected with the operation of the reactor. Furthermore, the Director continuously promotes safety culture through staff training and information, through implementation of effective internal communication and by encouraging personnel to suggest innovative solutions and areas for improvement.

— *Definition of working practices:* Working practices are documented in manuals, procedures and instructions, which constitute a guide for personnel to carry out safety related activities. Based on a graded approach, the documentation may have different levels of detail depending on the safety impact and complexity of the operations to be performed and on the competence of the staff; this approach ensures the reproducibility, traceability and control of activities.

— *Staff competence and training:* The capabilities of staff are a key factor in the implementation of the management system. The activities carried out are based on personnel skills and technical ability. The organization manages its personnel training according to the following criteria:
  • Definition of the staff skills necessary to carry out assigned activities affecting safety;
  • Identifying, planning, implementing and verifying training activities;
  • Ensuring that staff are aware of the importance of their activities and how they could affect safety;
  • Maintaining appropriate records of training, experience and skills acquired by personnel.

— *Audit and assessment:* The organization has developed an internal audit system applicable to all processes of the LENA QMS, including all the safety related activities aimed at satisfying mandatory conditions for both nuclear safety and health protection (for a more detailed description see Section I–3.9).

— Senior management commitment is realized through:
  • Establishing a quality policy and raising consciousness, motivation and involvement among the staff;
  • Setting safety objectives;

- Systematically verifying the satisfaction of requirements;
- Raising awareness in the organization about the importance of satisfying safety requirements;
- Carrying out management reviews;
- Ensuring the availability of resources;
- Systematically monitoring management system improvements.

I–2.2. Decision to develop a quality management system (QMS)

A prerequisite for the management of a research reactor is the satisfaction of all the requirements of interested parties, including safety constraints, security, environmental and occupational health requirements and the efficiency and effectiveness of the delivery of those services.

In order to continuously improve the quality of reactor management and the accomplishment of the requirements of interested parties, LENA decided to implement an integrated management system in accordance with The Management System for Facilities and Activities [I–5] and the International Quality Standard ISO 9001 [I–11] , among others.

This choice allowed LENA to satisfy both compulsory national and international requirements (i.e. safe reactor operation and maintenance) and typical ISO 9001 requirements (such as continuous improvement, care of users and interested parties and attention to their satisfaction).

Through this systematic and graded approach that led to the standardization of all processes involved in reactor operation and maintenance, all the aspects of the reactor management mentioned in Ref. [I–5] were satisfied.

In fact, the above mentioned reference provides a guidance system for establishing, implementing, assessing and continually improving a management system for facilities and activities that integrates safety, health, environmental, security, quality and economic elements.

This Annex and in particular the examples in Section 3 aim to explain how these recommendations can be implemented and developed by an medium sized organization managing a research reactor.

I–2.3. Analysis of processes involved in reactor operation and maintenance

One of the first steps in the setup of the management system was the specific and the detailed analysis of all the processes involved in the management of the organization activities, with particular emphasis on reactor safety operation and maintenance aspects.

The term 'process' refers to an integrated set of activities that uses resources to transform inputs into outputs, as shown in Fig. I–5. From a general point of view, the outputs from a process may become inputs for one or more different processes, therefore whenever several processes are interconnected, a system can be created through such input and output relationships.

According to this definition, LENA has developed its system through a 'process approach', which can be understood as a management strategy aiming to oversee, actuate and monitor the interactions between these processes, along with their interfaces with the functional hierarchies of the organization.

In line with this strategy, the number and types of processes needed to achieve the objectives were defined. In the LENA QMS, processes can be grouped as shown in Table I–5.

The requirements of interested parties, which are considered input elements, play a key role in the LENA QMS since they are determined both by thenational and international legislation applicable in the field of management of a nuclear facility as well as by the users (typical commercial requirements on service provision, i.e. ISO 9001) (See Sections I–1.3 and I–1.4).

For this reason, the main purpose of the implementation of the LENA QMS is the achievement of safe reactor operation through the control of a primary process and its correlated support processes. Therefore, based on the analysis of the requirements, the processes needed for achieving the above mentioned objectives were accurately defined.

TABLE I–5. PROCESS GROUPS AND DEFINITIONS

Process group	Definition
Organization's management processes	Processes relating to strategic planning, establishing policies, setting objectives, ensuring communication, ensuring availability of resources for quality objectives, for the desired results of activities and for management review
Resource management processes	Processes necessary to provide the resources needed to achieve the organization's objectives and desired outcomes
Realization processes	Processes required to perform activities, to ensure the fulfilment of objectives and requirements
Measurement, analysis and improvement processes	These include the processes needed to measure and gather data for performance analysis and improvement of effectiveness and efficiency. They include measuring, monitoring, auditing, performance analysis and improvement processes.

For a better development of the system, LENA defined as a primary process the 'operation of the TRIGA MARK II research nuclear reactor for the design and delivery of irradiation services' as a starting point from which to build the system, with a focus on compliance with specific requirements and also encompassing all other processes involving safety.

All the processes are described in specific documents, such as documented procedures, work instructions, manuals, etc., which have to be considered, along with their records, as an unavoidable process control tool. A schematic diagram of that described above as well as how processes are linked together with safety aspects is shown in Fig. I–6.

Primary process definition. The process that is usually defined as 'primary' is the process that constitutes the core business and first purpose of an organization. For a research reactor, this typically consists of research, irradiation and experimental activities.

Hence, LENA defined as its primary process the operation of the TRIGA MARK II research nuclear reactor for design and delivery of irradiation services.

To provide irradiation services and/or perform experiments at a research reactor implies that the requirements for safe reactor operation, including maintenance activities, have to be covered and complied with. So understood, the primary process includes all stages of design, planning and delivery of irradiated samples and/or experiments at the reactor.

Following this approach, the organization planned and developed the processes needed for its service provision, ensuring that the realization was consistent with the applicable requirements (see Section I–1.3).

During planning operations, the following issues were defined:

— Safety and quality objectives and requirements for the operation;
— The need to establish processes and to provide documents and resources specific to each process;
— Verification, validation, monitoring, measurement, inspection and test activities specific to the irradiation process and the criteria for product acceptance;
— Records needed to provide evidence that the realization process and its results meet the planned requirements.

During the stages of the realization of irradiation experiments, these needs were also defined:

— Design and development needs;
— Review, verification and validation appropriate to each design and development stage;

— Responsibilities and authorities for design and development;
— Management of the interface between different groups involved in design and development in order to ensure effective communication and clear assignment of responsibility.

As a result, irradiations and experiments are continuously carried out under controlled conditions that include the following as parts of the primary process:

— Definition of additional user requirements;
— Communication with users and interested parties;
— Inquiry and offer handling;
— Sample acceptance check;
— Irradiation design and implementation (standard and ad hoc experiments);
— Authorizations;
— Irradiation process;
— Process conformity check;
— Sample and irradiated products storage;
— Final checks for returning irradiated sample to users or managing irradiated material as radioactive waste;
— Return of irradiated samples to the user or client.

All the above topics are detailed in specific documented procedures and work instructions in order to ensure the desired efficiency and effectiveness of the process.

Support processes: As mentioned above, there are many other processes related to the primary process, which are usually referred to as 'support processes'. These processes are not part of the primary process, but they are essential in order for the primary process to take place.

In the case of a research reactor, all activities necessary to provide irradiation and/or experiments requiring reactor operation constitute what are known as support processes — all the activities connected with safe reactor operation and maintenance.

Support processes have been analysed, developed and implemented according with the process approach and can be grouped as shown in Table I–6.

Following the analysis of their processes, LENA also defined an overall process map. It plays a very important role in understanding and managing the whole system and shows both primary and support processes with their interconnections. Fig. I–7 shows the processes interaction diagram as reported in the LENA QMS manual.

TABLE I–6. OVERVIEW OF LENA SUPPORT PROCESSES

Process	Application field/remarks

Safety related processes

Process	Application field/remarks
Health physics surveillance	• Area radiation monitoring (continuous monitoring); • Daily radiation monitoring (in standby conditions); • Surface contamination control; • Personal dosimetry; • Personal contamination control; • Internal contamination control; • Radiation monitoring during irradiation; • Sample radiation monitoring; • Monitoring of irradiated products.
Environmental and waste monitoring	• Radioactive liquid effluents management; • Radioactive liquid waste management; • Radiation monitoring of building outlet air; • Air particulate activity monitoring; • Leakage monitoring; • Environmental dosimetry.
Maintenance	Mechanical • Reactor heating, ventilation and air conditioning system; • Reactor cooling system; • Inlet water treatment; • Fire detection and extinguisher device system; • Reactor building air filtering system; • Diesel generator. Electrical • Power supply system; • Reactor control instrumentation; • Public address and evacuation systems. Instrumentation • Reactor control instrumentation; • Area monitoring instrumentation; • Radioprotection devices and instrumentation; • Measurement equipment calibration and maintenance; • Security systems.

TABLE I–6. OVERVIEW OF LENA SUPPORT PROCESSES (cont.)

Process	Application field/remarks
Purchasing	• Purchasing of goods, materials and services related to the safety operation and maintenance of the reactor; • Safety equipment for workers.
Abnormal events Accident and/or emergency	• Internal and external incident reporting; • Internal emergency instructions; • External emergency plan; • Emergency equipment controls; • Periodic nuclear emergency drill.

Management processes

Organization's management processes	• Definition of responsibilities and authorities; • Development and implementation of the LENA QMS; • Continuous improvement of safe reactor operation; • Review of quality management system at planned intervals; • Communication to promote effective management of the quality management system.
Work environment control	• Conventional occupational HSE aspects; • General purpose power supply; • Lighting system; • Emergency lighting; • Security systems.
Control of documents	• Management (preparation, verification, approval, distribution, reviewing, archiving); • Documented procedures; • Work instruction; • Diagrams, manuals. • Control of documents refers to both internal and external documents (i.e. laws, regulations, etc.).
Control of records	All the records necessary to give evidence of procedures and actuations.
Resource management	Processes necessary to provide adequate resources for the organization's objectives and desired outcomes, including worker qualification and training.

TABLE I–6. OVERVIEW OF LENA SUPPORT PROCESSES (cont.)

Process	Application field/remarks
Internal and external communication	Processes ensuring that appropriate communications are established within the organization and that communication takes place regarding the effectiveness of the Management System. Processes for both internal and external communications are necessary.
Internal audit	An audit programme is planned, taking into consideration the status and importance of the processes and areas to be audited, as well as the results of previous audits. The audit criteria, scope, frequency and methods are defined.
Control of non-conformances	Products which do not conform to product requirements are identified and controlled to prevent their unintended use or delivery. A documented procedure is established to define the controls and related responsibilities and authorities for dealing with non-conforming products.
Corrective action (CA) and preventive action (PA)	Action to eliminate the cause of a detected non-conformance in order to prevent recurrence (CA) or action to eliminate the cause of a potential non-conformance (PA).
Monitoring and measurement of processes	Methods for monitoring and, where applicable, measuring the management system processes. These methods demonstrate the ability of the processes to achieve planned results. When planned results are not achieved, corrective action is taken, if appropriate.

I–2.4. Implementation strategy: the main steps leading to the QMS implementation at LENA

In order to accomplish the implementation of the LENA QMS with a rational logic and within the scheduled time, the entire management system was designed in accordance with a General Plan for Design and Development of the Quality Management System drawn up by the University of Pavia, which defined a specific and precise strategy for the set-up of the system. The outline in Fig. I–8 shows the step by step path followed for the implementation process.

TABLE I–7. FORMER DOCUMENTAL SYSTEM

Existing documental system	Document title
Prescriptive documents	Licence for Reactor Operation
	Operating Rules
	Management Rules (Operational Conditions)
	Operational Requirements (Operational Limits)
	Plant Supervisory Rules
	Emergency Plan
Operating instructions	Operational Instructions
	Technical Specifications for Plant Nuclear Tests
	Health Physics Regulation
	Reactor Maintenance Procedures
	Internal Emergency Instruction
Other relevant documents	Final Safety Assessment
	Power Outage Analysis Report
	Plant Nuclear Tests Reports
	Plant Operation and Maintenance Report
	Fire Protection Devices and Fire Procedures
	Service Orders

I–3. EXAMPLES OF MANAGEMENT SYSTEM APPLICATION

I–3.1. Initial status

As a nuclear reactor facility, the organization has to deal with a complex documentation system. Prior to the LENA QMS implementation, the documentation utilized by staff for reactor operation and management could be summarized (hierarchically) as in Table I–7.

That documentation system was, of course, related to both national and international legal requirements (see Section I–1.4), so the above table and the related documents were subject to change according to other regional and local requirements.

Therefore, as far as the implementation of the management system was concerned, one of the first tasks was an inventory of the existing document system and an analysis of how the LENA QMS could build on and integrate it.The implementation process, in order to fulfil its purpose, as described in the previous sections, required the development of new documents that could integrate, interact and complete the existing documental system. Figure I–9 shows the hierarchical dependency of the LENA documental system after the LENA QMS implementation.

The implementation efforts resulted in the development of approximately 50 new documents and the revision of approximately 20 existing documents. These consisted of:

— LENA QMS manual;
— Quality procedures;
— Technical procedures;
— Maintenance procedures and instructions not otherwise documented;
— Work instructions;
— Forms (in addition to pre-existing forms).

It is interesting to underline how the development and application of the new documentation did not in any way encumber the management of the system, but rather improved it both in terms of efficiency and effectiveness, covering all the aspects not yet formally considered. Table I–8 contains a non-exhaustive list of the documentation LENA produced to complete the LENA QMS.

I–3.2. Quality management system development

The design and development of the LENA QMS was planned in detail and implemented according to the steps already described in I–2.4. The project team identified the main tasks to be completed and their priority. The main topics taken into account for the project development were:

— Task list definition;
— Staff involved in each task;
— Task interaction;
— Priority list;
— Time schedule;
— Milestone definition;
— Delay management.

In order to deal with the above mentioned issues, the well known Gantt chart approach was chosen as a project management tool.

A Gantt chart is a bar chart that illustrates a project schedule. In order to define the project work breakdown, it illustrates the start and finish dates of the individual activities, their connections and the summary elements of a

TABLE I–8. DOCUMENTATION PRODUCED BY THE ORGANIZATION
TO COMPLETE THE MANAGEMENT SYSTEM

DOCUMENTATION ISSUED DURING MANAGEMENT SYSTEMS DEVELOPMENT
Irradiation Process Design and Management (Procedure)
Management Review (Procedure)
External Communication (Procedure)
Documentation Management (Procedure)
Records Management (Procedure)
Staff Training (Procedure)
Equipment Management (Procedure)
Mechanical Maintenance (Plan and instructions)
Electrical and Instrumentation Maintenance (Plan and instructions)
Health Physics (Instructions)
Purchasing Management (Procedure)
Internal Audits (Procedure)
Non-conformance Management (Procedure)
Corrective and Preventive Actions (Procedure)
User Satisfaction/Feedback (Procedure)
Process Measurements (Plan and records)
Management System Manual

project. The Gantt chart also shows the dependence (i.e. precedence network) relationships between activities.

This tool allows all people involved in the project to be continuously updated on the current progress status; in fact, it shows the current activity progress as a percentage shown by the shading of the activity bar and a vertical 'today' line as an indicator of the complete work progress as of the current date.

Figure I–10 gives an example where the following can be identified:

— The time line bar (in the upper part), showing the foreseen dates of the ongoing activities;

— The name of the activity considered (e.g. quality policy editing and arrangements) and the foreseen duration (progress bar);
— On the right side of the progress bar the personnel dedicated to that activity are identified (staff assigned).

The specific use of this tool helped accomplish the project within the scheduled time. The following aspects have to be taken into account:

— *Planning the project.* A commercially available software tool was used to define the different tasks and to determine the critical path of the project. A critical path shows the longest path of planned activities from the beginning to the end of the project, as well any interdependence between activities, allowing the calculation of the earliest and latest date when each activity can start and finish without lengthening the project timeline. The software also shows the influence of a delay in any activity on the overall project schedule. LENA decided to focus on the activies in the critical path first, to avoid accumulating delay in the overall project process. As a result of this approach, the implementation process ended within the scheduled time, which was, from start to finish, approximately two years.
— *Day-to-day management of the project.* During the project the Gantt chart was continuously updated. This required a clear internal communication process in order to receive the necessary feedback from the staff invovled in the scheduled tasks, such as holding team meetings, periodically verifying the project progress status, etc. This process has been shown to be successful.

Figure I–10 shows an extract from a LENA chart referring to the final part of the implementation process, showing tasks, task relationship, staff assignments, duration, delays, etc. A more detailed view is presented in Figs I–11 and I–12.

Figure I–12 shows a LENA activity schedule where the planned duration, the current completion status and the assigned resources are presented for each activity.

I–3.3. Management System Manual

The LENA QMS meets the requirements of all interested parties along with those of the International Standard ISO 9001:2008. The MS Manual is a document that provides an overview of how the organization and its management system are designed to meet its policies and objectives. The manual describes the LENA QMS and its purpose and it delineates the authorities, interrelationships and responsibilities of the personnel within the system. The manual also provides procedures or references for all activities that comprise the LENA QMS to ensure compliance with the necessary requirements.

The manual is used internally to guide the organization's employees through the various requirements that must be met and maintained in order to ensure safe operation, user satisfaction and continuous improvement, providing the necessary instructions to create an empowered work force.

The manual can be used externally to introduce the organization's management system to all interested parties and other external organizations or individuals. The manual is used to familiarize them with the controls that have been implemented and to assure them that the integrity of the management system is maintained and that it focuses on the satisfaction of requirements, on safe operation, user satisfaction and continuous improvement.

The manual is divided into eight sections. Each section provides specific information regarding the procedures that describe the methods used to implement the system and meet the necessary requirements. Table I–9 shows the LENA Management System Manual in relation to the applicable requirements.

TABLE I–9. CONTENT OF THE LENA MANAGEMENT SYSTEM MANUAL IN RELATION TO APPLICABLE REQUIREMENTS

Management system manual	Topics, related procedures or instructions
Section 0 Forewords	LENA profile and presentation
Section 1 Scope and field of application	Illustrates the scope and field of application of the QMS
Section 2 Normative reference	Describes the applicable requirements and standards (i.e. national laws and international regulations, guidelines and international standards (see Section I–1.4))
Section 3 Terms and definitions	Describes terms, definitions and acronyms adopted in the LENA QMS
Section 4 Management system	Includes, among other things, a description of the following (or reference to other specific documents): • Management system manual; • Process descriptions; • Process interactions; • Control of documents; • Control of records.

TABLE I–9. CONTENT OF THE LENA MANAGEMENT SYSTEM MANUAL IN RELATION TO APPLICABLE REQUIREMENTS (cont.)

Management system manual	Topics, related procedures or instructions
Section 5 Management responsibility	Includes, among other things, a description of the following (or reference to other specific documents): • Senior management commitment; • Focus on users and interested parties; • Quality policy; • System planning; • Responsibility and authority; • Internal communication; • Management review.
Section 6 Resource management	Includes, among others, a description of the following (or reference to other specific documents): • Provision of resources; • Human resources; • Competence, awareness and training; • Infrastructure; • Work environment (maintenance and HSE).
Section 7 Service provision	Includes, among others things, a description of the following (or reference to other specific documents): • Service provision planning; • Customer related processes; • Design and development; • Purchasing; • Control of production and service provision; • Validation of processes for production and service provision; • Control of monitoring and measuring equipment.
Section 8 Measurement, analysis and improvement	Includes, among other things, a description of the following (or reference to other specific documents): • Customer satisfaction; • Internal audit; • Monitoring and measurement of processes; • Monitoring and measurement of services; • Control of non-conforming products; • Analysis of data; • Continual improvement; • Corrective action; • Preventive action.

I–3.4. Operation and maintenance management

With the terms 'operations and maintenance', the organization encompasses the broad spectrum of services and activities required to ensure that the reactor will perform the functions for which it was designed and constructed. Operations and maintenance typically include the day-to-day activities necessary for the reactor, its systems and equipment to perform their intended functions and maintain them during their life cycle. These activities include preventive (planned) maintenance and corrective (repair) maintenance.

Preventive maintenance (PM) consists of a series of time based maintenance requirements that provide a basis for planning, scheduling and executing planned maintenance. As far as the reactor safety operation is concerned, most PM activities, as well as their scheduling times, are mandatory since they are determined by the legal requirements applying to the organization for reactor operation (See Section I–1.4). PM includes, for example, testing, adjusting, lubricating, cleaning and replacing components. Time intensive PM, such as bearing or seal replacement, or instrumentation and control system calibration, would typically be scheduled for regular reactor shutdown periods.

Corrective maintenance (CM) is repair needed to return the equipment to a properly functioning condition; it may be planned or unplanned. CM also includes, where applicable, instruments repairs and system revamping or upgrading.

Both PM and CM are managed through specific internal procedures or work instructions, always with traceability of the activities and the correlated support documentation.

Operations and maintenance interacts with the LENA QMS through the maintenance programme (1 year cycle) and, vice versa; the LENA QMS interacts with operations and maintenance through non-conformance (NC), corrective action (CA) and preventive action (PA) management.

Operations and maintenance efficiency and effectiveness is monitored through the definition of specific performance parameters with the intention of continual improvement. Operations and maintenance requires a knowledgeable, skilled and well trained management and technical staff. The goals of a comprehensive maintenance programme include the following:

— Ensuring the safe operation of all equipment;
— Providing safe, functional systems and facilities that meet the intended design;
— Reducing unscheduled shutdowns and repairs;

TABLE I–10. OVERVIEW OF LENA OPERATIONS AND MECHANICAL MAINTENANCE PLANNING

System	Activity	LENA reference procedure or instruction	Frequency	LENA form identification code
Primary cooling system	Primary pumps check	IL SMM 6.3/04	15 days	MD SMMCQ
Ventilation system	HEPA filter replacement	IL SMM 6.3/24	1 year	RMM
	Air flow rate measurement	PS 4.2.4	15 days	MD SMMCQ
	Active charcoal filter replacement	IL SMM 6.3/25	18 months	RMM
Control rods	Rod lubricating and sealing verification	IL SMM 6.3/18	6 months	RMM
Compressed air circuits	Pressure gauges calibration/ adjustment	IL SMM 6.3/28	6 months	MD SMM 6.3/28
Fire detection system	Device check and cleaning	Outsourced	6 months	RA
Generating set	Startup checks, oil and refrigerant levels, battery check	IL SMM 6.3/14 IL SMM 6.3/15	15 days	MD SMMCQ

— Extending equipment life, thereby extending facility life;
— Realizing life cycle cost savings;
— Reducing capital repairs.

Tables I–10 and I–11 represent an example of the operations and maintenance plans for both mechanical and electrical reactor maintenance (general building operations and maintenance programmes are neither shown nor discussed here).

Work instruction example. A part of a work instruction is presented in Fig. I–13 in order to show the typical structure of the instructions used.

TABLE I–11. OVERVIEW OF LENA ELECTRICAL OPERATION AND MAINTENANCE PLANNING

System	Activity	LENA reference procedure or instruction	Frequency	LENA form identification code
Reactor control instrumentation	Ion chambers HV power supply voltage readings	PS 4.5.1.1.E	6 months	MD SMESICS
	Ion chambers resistance measurements	PS 4.5.1.1.D	6 months	MD SMESICS
	Strip chart linear and log recorder adjustment and calibration	PS 4.5.1.1.B	6 months	MD SMESICS
	Fuel temperature channel Meter calibration and thermocouples verification	PS 4.5.1.1.F	6 months	MD SMESICS
Portable radioprotection equipment	Calibration test	PS 4.4.3	6 months	RSF
Area monitoring system	Calibration test	PS 4.4.1B	3 months	RSF
Reactor power supply system	RCD check and trip test	MD ME 6.3a	1 month	RME

I–3.5. Health physics surveillance, environmental monitoring and waste management

The Health Physics Section employs highly qualified staff and operates a sophisticated system of monitoring instruments, radiochemical laboratories, networks of local environmental radioactivity monitoring and mobile devices to cope with any abnormal situations that could have consequences for the external environment.

Since health physics surveillance is of key importance to ensure the safe operation of the reactor, the organization has implemented and continuously applies a well defined control plan for health physics and environmental protection. The control plan, as defined by legal and regulatory requirements, by specific internal regulation as well as other legal and regulatory requirements, is implemented through a series of activities and checks carried out at periodic intervals. Figure I–14 shows the structure of health physics surveillance activities.

Table I–12 shows a short, more detailed description of the key health physics activities to ensure the protection of workers, the general population and the environment, and checks carried out on waste, as well as frequencies, which relate to an established minimum period. In the event of different necessities, frequencies may change.

TABLE I–12. KEY HEALTH PHYSICS CHECKS CARRIED OUT ON WORKERS, ENVIRONMENT, POPULATION AND WASTE

HEALTH PHYSICS SURVEILLANCE PLAN				
Control	Workers and population	Environmental	Waste	Frequency
Area radiation monitoring (inside reactor building)	•			Continuous data acquisition
Personal dosimetry	•			Monthly personal dosimeter readings
Personal contamination control	•			Always when leaving controlled areas
Surface contamination control	•		•	1 month
Emergency equipment controls	•	•		2 weeks
Internal contamination control (WBC/gamma spectrometry biological sample)	•			6 months
Guest dosimetry	•			During and after each visit to controlled areas
Personal dosimetry report	•			1 yearly

HEALTH PHYSICS SURVEILLANCE PLAN

Radiation monitoring (reactor in shutdown condition)	•	•		Daily
Radiation monitoring during operation	•	•		During and after each reactor operation
Irradiated products packaging and handover controls	•			For each sample before delivery
Liquid effluents storage	•	•		1 month and before each drain release
Radioactive liquid waste	•	•	•	Before move to warehouse
Air effluents	•	•	•	Continuous data acquisition
Air particulate activity monitoring	•	•		Daily
Fission products monitoring (tank water spectrometry)	•			1 month
Environmental dosimetry	•	•		1 month
Solid waste		•	•	Before moving to storage facility
Solid waste warehouse		•	•	6 months
Ar-41 air release evaluation report		•	•	1 year

In abnormal conditions, the mentioned plan is integrated with supplementary checks in order to ensure the safety of the population and of emergency and rescue staff involved in emergency management.

As far as non-radioactive HSE (health, safety and environment) aspects are concerned, the organization manages waste, environment and workers' safety according to national laws and regulations.

I–3.6. Realization, review, approval, manufacturing and commissioning of experiments

The main activity carried out by LENA is the operation of the reactor for the irradiation of materials and products as part of experiments conducted by the organization itself or by other users (researchers, companies, etc.).

The irradiation process (primary process, see Section I–2.4), as well as the entire LENA QMS, was designed in accordance with the LENA document General Plan for Design and Development of the Management System.

Nevertheless, if it becomes necessary to plan and realize specific activities not covered by the standard irradiation procedures (due to changes in an already ongoing experiment or because of the planning of new experiments) appropriate steps and responsibilities are defined, as shown in Fig. I–15.

The design activities are kept up to date through the drafting of meeting reports, through the gradual completion of the monitoring design development form and through the issue of procedures and/or their review.

Realization input elements are identified in the initial feasibility analysis and may include:

— All statutory and regulatory requirements applicable to the operation of a nuclear installation and to the management of radioactive material;
— Historical information, including records, from previous operational experience and experiments at the reactor;
— Functional requirements, e.g. physics and control of nuclear installations, as well as requirements for facilities and equipment that can influence the quality of operation or irradiation;
— Procedures established for the implementation of activities;
— The competence required by personnel assigned for specific activities within the operation or irradiation process;
— Implicit requirements;
— User requirements, depending on the different kinds of users;
— Available resources;
— Quality objectives of the process or activity under consideration.

All requirements are periodically reviewed in order to verify adequacy, completeness and absence of ambiguity and conflict.

The outputs of the realization process are procedures, work instructions, and any other documentation which illustrates the management and operating methods used by personnel for the conduct of activities.

The outputs of the design include, or refer to, if appropriate, the following elements:

— Information for the supply of goods and services, if required;
— Criteria and specifications for compliance with the process and for products;
— Safety features of the processes;
— Specifications for the storage and delivery of irradiated products.

The verification of design is peformed to ensure that the specifications have been met and transposed into the outputs of the design. It consists of checking the completeness and accuracy of data and the achievement of expected results.

In accordance with the plan, the design review is implemented at appropriate stages (previously determined and scheduled based on the complexity of the project). It allows assessment of whether the results meet the design input requirements and identification of any shortcomings and necessary actions (i.e. design changes). The review is performed with the participation of the functions involved in planning and development (e.g. through group meetings) and of the Director (e.g. in an extra management review).

At the end of the verification and review phases, the project validation is carried out in order to determine compliance with requirements. The experiment validation is performed on the first irradiation or experiment or is based on historical data if available.

The project validation is intended to confirm that:

— The designed experiment satisfies the needs and requirements identified in the design phase.
— The realization process works in accordance with the newly implemented procedures.
— Assigned resources (personnel and equipment) are adequate, appropriate and actually available.
— Information provided to the user is complete and comprehensive.
— The designed process can effectively meet the requirements, achieve the quality objectives and the planned results.

If design changes are proposed and evaluated before the phases of verification, validation and final review, they are considered and handled in these phases. If changes are introduced after these steps, it is necessary to begin again with verification, review and validation before the delivery of the new service is implemented.

I–3.7. Management review

In order to ensure continuous suitability, adequacy and effectiveness, the LENA QMS and its implementation is reviewed at planned intervals by LENA's Director, based on information gathered from the Quality Assurance Unit.

The review includes evaluations for improvement opportunities and for changes needed in the LENA QMS, including quality policy and quality objectives. Records from management reviews are maintained and utilized as input for further management review meetings.

The organization applies a specific documented procedure for the conduct of management reviews.

The input to management review includes information concerning:

— Results of internal audits;
— NC reports;
— Status of preventive and corrective actions;
— Follow-up actions from previous management reviews;
— Process indicator evaluation reports;
— Changes that could affect the LENA QMS and recommendations for improvement;
— LENA QMS budget review;
— Operational feedback;
— Incident and accident reports;
— User feedback.

As a result, the output from the management review includes any decisions and actions related to:

— Improvement of the effectiveness of the LENA QMS and its processes;
— Improvement plans;
— New expected values for process indicator;
— New preventive (or corrective) action;
— Resource needs;
— Improvement of services related to user requirements.

Figure I–16 illustrates the key steps and interactions of the LENA management review process:

I–3.8. Non-conformance, preventive action and corrective action management

The organization ensures that processes and products that do not conform to requirements are identified and controlled to prevent unintended use or delivery. A documented procedure is established to define the controls, related responsibilities and authorities for dealing with non-conforming processes or products.

Owing to its particular operation field, LENA identified two different types of non-conformance (NC):

— NC: non-fulfilment of a requirement, as defined in the main international quality standards (e.g. NC related to inappropriate sample storage before irradiation);
— Safety related NC: non-fulfilment of requirements that may affect safe reactor operation or health physics aspects (e.g. NC regarding reactor control system instrumentation, incidents and unusual events).

In the case of a detected non-conforming product, one of the following actions might be taken:

— Elimination of the detected NC (correction);
— Authorizing its use, release or acceptance under concession by a relevant authority and, where applicable, by the customer;
— Taking action to preclude its original intended use or application;
— Informing the user with a proposal for correction when an NC is detected after the irradiation/experiment has been concluded and the samples have already been shipped.

When a non-conforming product is corrected, the product or service is subjected to re-verification to demonstrate conformance to the requirements.

Records on the nature of any NC and any subsequent actions taken, including concessions obtained and correction costs, are maintained.

Corrective Actions (CA). A CA is an action to eliminate the cause of a detected NC or other undesirable situation. CAs must be appropriate to the effects of the detected NCs. A documented procedure is established to define requirements for:

— Reviewing NCs (including complaints);
— Determining the causes of NCs;
— Evaluating the need for action to ensure that NCs do not reoccur;

— Determining and implementing action needed;
— Records of the results of action taken;
— Reviewing the effectiveness of the corrective action taken.

Preventive action (PA). This action aims to eliminate the causes of potential NCs in order to prevent their reoccurrence. Preventive actions are appropriate to the effects of the potential problems. The organization developed and applies a documented procedure to define requirements for:

— Determining potential non-conformances and their causes;
— Evaluating the need for action to prevent occurrence of non-conformances;
— Determining and implementing action needed;
— Recording results of action taken and reviewing the effectiveness of the preventive action taken.

As can be seen from the flow chart in Fig. I–17, the management of safety related NCs implies the utilization of a different and more accurate procedure in order to eliminate the NC. LENA has implemented specific procedures and forms to define the authority, responsibility and operating instructions connected with safety operation activities. In some cases, for better management of safety related NCs, further support documentation may be required.

I–3.9. Internal audits

In the context of a management system, measuring the effectiveness of the processes and the achievement of objectives is essential. This process must be properly planned to identify and apply methods of process performance measurement and ensure that it is possible to systematically assess:

— The compliance of the LENA QMS with rules, laws and regulations applicable to quality and strategically relevant factors (such as nuclear safety and health protection);
— The effectiveness of the processes included in the LENA QMS.

LENA has adopted appropriate methods to monitor and, where applicable, measure LENA QMS processes, in order to demonstrate their ability to effectively provide the planned results and to ensure adequate performance in the areas of the LENA QMS. One of these methods is their internal audit system. Within the LENA QMS, the procedure PTGQ8.2/2, Internal Audit, has been implemented. It describes the operating procedures, responsibilities, documentation and functions involved in planning, scheduling and conducting internal audits. The process is shown in Fig. I–18.

In particular, this procedure describes activities for:

— Planning and scheduling audits;
— Conducting audits;
— Presentation and recording of results including deviations detected or opportunities for improvement;
— Preservation of produced records.

In accordance with regulatory standards, the procedure is intended to:

— Ensure that LENA QMS activities and their results meet set objectives and comply with requirements as well as with applicable rules;
— Assess the implementation and effectiveness of the LENA QMS;
— Ensure that the quality policy, the quality manual requirements and the LENA QMS plans and procedures are effectively applied and, if necessary, enforced.

The internal audit system is applicable to all processes of the LENA QMS, including all the activities planned for the purposes of complying with mandatory conditions for nuclear safety and health protection. The unit responsible for the management of internal audits is the Quality Assurance Unit, which issues the audit plans and programmes, collects documents relating to internal audits, agrees to corrective actions and improvement with involved staff and prepares audit results for management review.

The audit programme takes into consideration all elements of the QMS at least once in an annual cycle of inspections. During an audit, it is possible to check:

— The whole LENA QMS;
— One or more elements of the LENA QMS in detail (e.g. control documentation, control of records, implementation of specific controls/procedures/instructions);
— The correct application of the measures and procedures related to nuclear safety and health protection requirements.

The competence of an auditor is be based on the following criteria:

(a) Knowledge of the audited processes and activities;
(b) Knowledge of LENA QMS standards;

(c)   Experience in managing and conducting audits.

The choice of the members of the audit team must be such that:

— All the knowledge and skills required by the above criteria are present in the audit team;
— The objectivity and impartiality of the audit team is ensured.

The audit records (such as checklists and audit reports) are maintained and reported to the Director and to relevant plant sections, for the purposes of subsequent evaluation and actions, if required.

The action taken (such as CA or PA) are always considered during the next audit cycle (or in additional audits if determined to be necessary) and are included among the inputs to the Management Review.

I–3.10.   Performance indicators

Performance indicators are data that provide a measurement of some aspect related to process efficiency and effectiveness. They are an indispensable tool to manage, assess and improve the LENA QMS. During the implementation process, for each process involved (see Section I–2.3) the organization identified a set of parameters to benchmark the process itself as well as the whole system. Each indicator can be represented as a numerical calculation of a specific formula. Input data are collected from the records from a predefined period of time onwards.

Calculated results can definitely be considered one of the most important inputs for a system or management review, in order to improve the system, to take actions to correct eventual LENA QMS gaps or lacks and for the continual improvement of the organization.

Indicators have to cover all aspects of the system. They may range, for example, from reactor safety operation to the satisfaction of interested parties.

As far as the indicators are concerned, the definition of their expected values was based on historical data already collected and, where data was not available, on reasonable values determined by the experienced project team.

Depending on the results of the measurements, LENA determined the proper actions or strategies in order to achieve the desired targets.

Table I–13 shows, as an example, a selection of indicators chosen to monitor and assess progress in reaching the desired objectives. The reference period considered for the calculations is one work year (2010).

In particular, it is interesting to observe values obtained for indicators related to reactor operation processes and how those indicators match up with

TABLE I–13. OVERVIEW OF PERFORMANCE INDICATORS CALCULATED BY LENA

Process	Indicator	Expected value	Calculated value
Reactor operation	Reactor availability*	70–80%	77%
	Reliability*	70–80%	89%
	Reactor utilization*	>50%	52%
Instrumentation management	Failure probability (no. of repairs/no. of instruments)	<15%	23%
	Calibration service efficiency (no. of instruments found out of calibration)	0	0
Human resources	Percentage of employees who attended at least one training course	>50%	100%
Purchasing	Purchasing verification efficiency (NC in phase of service provision/ NC detected during verification of purchased products + NC in phase of service provision)	→0	0
Maintenance	Mean time to repair, in days (MTTR)	5	3
	Outsourcing service dependency (no. of maintenance operations requiring outsourcing assistance/ no. of maintenance operations)	<15%	30%
LENA QMS	No. of safety related NCs detected	→0	12
	No. of PAs	>1	4
	Auditing system efficiency (no. of CAs as a result of an internal audit)	>1	5

* Calculated according to IAEA Nuclear Energy Series No. NP-T-5.4 [I–9]

indicators in the publication Optimization of Research Reactor Availability and Reliability: Recommended Practices, IAEA Nuclear Energy Series No. NP-T-5.4 [I–9]. This choice was intended to provide evidence of reactor operational and safety related processes, as well as of efficiency, reliability and availability.

The following performance indicators are taken from Ref. [I–9]:

Availability: The fraction of time for which a system is capable of fulfilling its intended purpose:

$$\frac{\text{actual operating time}}{\text{planned operating time}} \times 100$$

— *Reliability:* The probability that a system or component will meet its minimum performance requirements when called upon to do so:

$$\frac{\text{actual operating time}}{(\text{actual operating time} + \text{duration of unplanned shutdowns})} \times 100$$

— *Utilization:* A measure of a facility's planned operation:

$$\frac{\text{planned operating time}}{\text{total duration of reporting period}} \times 100$$

I–3.11. Cost analysis

It is interesting to dwell on some aspects of the costs of the system implementation. In fact, for a typical medium sized organization managing a research reactor, its limited financial capacity, combined with the objective difficulty in the development of an accurate cost analysis plan, can be a major factor in the decision on whether to implement a certificated integrated management system.

In addition, the activities related to the system implementation, due to the specific application field, cannot avoid a significant involvement of internal resources both in terms of specific skills required and time spent on system implementation.

Carrying out a preliminary cost analysis would help to plan expenditure related to the implementation process. In the case of LENA, the project time frame was estimated as two years and, as a result of a first cost analysis, LENA identified the following main cost categories:

— *Project team involvement* (in person-hours, for the duration of the project): For each unit of personnel involved, a gross number of hours to be dedicated

120

to the system implementation has been estimated. Of course, for the project team, this commitment predominated.

— *Demand for internal and external training for the project team and the staff involved:* Needs for basic and/or advanced training on management systems and on specific standards (e.g. GS-R-3 [I–5], ISO 9001 [I–11]) were identified for each person involved in the project, with the aim of ensuring the proper knowledge of topics and familiarizing people with the application of quality principles 'in the field'. For the project team (in particular the team leader), there was a need for external advanced courses in order to master all the aspects related to the implementation process. For the rest of the staff, at least general training in the application of the LENA QMS was planned. Training in specific activities (i.e. for the application of new procedures) was provided by the team leader to the rest of the staff. Costs due to consultancy services (outsourced) also have to be taken into account. In the case of LENA, a consultancy service was provided by the University of Pavia (through commitment of internal qualified staff) to periodically assess the implementation progress as a support for the team.

— *Need for new equipment acquisitions:* The purchase of new equipment was planned in order to fulfil all the requirements related to metrological confirmation aspects. This cost category is more relevant to the required equipment managed by the organization than to the implementation of the QMS. In LENA's case, a few electronic instruments and reference materials were purchased, with a relatively small impact on the project budget.

— *Equipment management and calibration costs:* Before the beginning of the QMS implementation, LENA was already managing most of its radiation protection instruments according to an instrument calibration plan; therefore, only new costs necessary to integrate metrological confirmation aspects were considered.

— *Cost due to system certification (external certification body):* If an organization aims to gain external certification of its management system at the end of the implementation process, this is a cost to be considered in advance. Therefore, certifying and maintaining the system is a cost in terms of assessments performed by the external certification body, periodic inspections (generally once a year) and issuance of certificates.

Throughout the entire project, cost control strategies were employed in order to compare actual costs with the initial cost analysis. During the implementation process, it was noticed that most of the costs could be attributed (in terms of hours worked) to staff involvement (i.e. issue of new procedures, team meetings, implementation of new procedures, etc.). Based on the preliminary cost analysis, all the other investments were planned periodically, according to current needs.

As can be seen in Fig. I–19, all other resultant costs were marginal in comparison with those related to staff involvement.

In order to give a proper representation of data, and to avoid giving absolute monetary values that may vary between different countries (for example, due to a different cost of work and living), the aspects mentioned above are represented as a percentage of the total outlay for the implementation path. In particular, staff involvement in the project has been compared with the total amount of work hours per year (in Italy, 1440 work hours per year). Also considered in the cost analysis are the wage differences related to the different staff skills and experience. All costs have been calculated as gross outlays for the organization, including, where applicable, VAT and taxes according to national fiscal laws.

Figure I–20 shows the details of staff involvement in comparison with the total outlay. It is important to focus on the fact that a proper utilization of internal resources requires an accurate assessment of how much effort each task would require. As already mentioned (see I–1.2), in smaller organizations there is often multiple allocation of responsibilities for each personnel unit. Therefore, it is essential to define a proper work time schedule assignment for each employee involved in order to ensure that project progress does not affect the reactor operation. Table I–14 represents the personnel commitment during the project, showing the amount of hours worked on the project in comparison with the total hours worked.

For the above mentioned reasons, it is clear that comparable organizations, although small, can implement an efficient management system with a relatively small monetary investment, if most of the tasks are carried out by staff.

I–3.12.   Correspondence between this publication and LENA's QMS

Table I–15 shows how LENA's QMS takes into consideration all aspects described in the main body of this publication, together with the main reference documents developed (or modified) in order to comply with all the issues required for system implementation.

Text cont. on p. 146.

TABLE I–14. PERSONNEL COMMITMENT IN THE PROJECT: HOURS WORKED ON THE PROJECT IN COMPARISON WITH TOTAL HOURS WORKED

LENA staff groups (no. of employees involved in the project)	Hours working towards system implementation January 2009– October 2010	Hours worked on the project as a percentage of total hours worked[a]
Team Leader — QA Head	1250	43.4%
Quality Assurance Staff (1)	300	10.4%
Health Physics Section (1)	300	10.4%
Electrical & Instrumentation Maintenance Section (1)	400	13.9%
Mechanical Maintenance Section (1)	300	10.4%
Top Management (1)	150	5.2%
Administrative Office (1)	80	2.8%
Other staff involved (3)	100	3.5%
Consultancy services (not LENA staff)	(100)	N/A
Total	2880	100%

[a] LENA staff annual working hours = 1440 h/y

TABLE I–15. CORRESPONDENCE BETWEEN THIS PUBLICATION AND LENA QMS

Section and title of main text		LENA QMS	
		Process	Documents
2	Management system	Quality Management System	• All documents belonging to LENA's Quality Management System.
3	Management responsibility	Management responsibility	• Management System Manual.
3.1	Management commitment	Management commitment	• Management System Manual.
3.2	Management processes	Quality policy	• Management System Manual;
3.2.2	Planning and reporting	Planning	• LENA quality policy;
3.2.3	Management system administration	Responsibility, authority and communication	• Management System Manual; • Procedure 'management review';
3.2.4	Organizational structure	Management review	• Management review records; • Improvement plan.
4	Resource management	Human resources	• Management System Manual;
4.1	Recruitment, selection and appointment	General	• Procedure 'employee training';
4.2	Performance, assessment and training	Competence, training and awareness	• Training records;
4.3	Employing temporary workers or outsourcing		• Staff record.

TABLE I–15. CORRESPONDENCE BETWEEN THIS PUBLICATION AND LENA QMS (cont.)

Section and title of main text			LENA QMS	
		Process		Documents
Purchasing	5.1.5	Purchasing process	7.4.1	• Procedure 'purchasing management'; • Supplier evaluation form; • Supplier questionnaire.
Communication	5.1.6	Internal communication	5.5.3	• Management System Manual.
		Customer communication	7.2.3	• Procedure 'customer communications'; • Complaint forms.
Project management	5.2.1	Design and development Control of design and development changes	7.3 7.3.7	• Management System Manual.
Preparation of technical reports Incoming and outgoing correspondence Archiving	5.2.3 5.2.5 5.2.6	Control of documents	4.2.3	• Procedure 'documents management'; • Records list; • Completed forms and registrations.
Numerical calculations	5.2.4	Design and development	7.3	• Management System Manual.

TABLE I–15. CORRESPONDENCE BETWEEN THIS PUBLICATION AND LENA QMS (cont.)

Section and title of main text		LENA QMS	
		Process	Documents
2	Management system	Quality Management System	• All documents belonging to LENA's Quality Management System.
3	Management responsibility	Management responsibility	• Management System Manual.
3.1	Management commitment	Management commitment	• Management System Manual.
3.2	Management processes	Quality policy	• Management System Manual;
3.2.2	Planning and reporting	Planning	• LENA quality policy;
3.2.3	Management system administration	Responsibility, authority and communication	• Management System Manual; • Procedure 'management review';
3.2.4	Organizational structure	Management review	• Management review records; • Improvement plan.
4	Resource management	Human resources	• Management System Manual;
4.1	Recruitment, selection and appointment	General	• Procedure 'employee training';
4.2	Performance, assessment and training	Competence, training and awareness	• Training records;
4.3	Employing temporary workers or outsourcing		• Staff record.

126

TABLE I–15. CORRESPONDENCE BETWEEN THIS PUBLICATION AND LENA QMS (cont.)

Section and title of main text		LENA QMS	
		Process	Documents
2	Management system	Quality Management System	• All documents belonging to LENA's Quality Management System.
3	Management responsibility	Management responsibility	• Management System Manual.
3.1	Management commitment	Management commitment	• Management System Manual.
3.2	Management processes	Quality policy	• Management System Manual;
3.2.2	Planning and reporting	Planning	• LENA quality policy;
3.2.3	Management system administration	Responsibility, authority and communication	• Management System Manual; • Procedure 'management review';
3.2.4	Organizational structure	Management review	• Management review records; • Improvement plan.
4	Resource management	Human resources	• Management System Manual;
4.1	Recruitment, selection and appointment	General	• Procedure 'employee training';
4.2	Performance, assessment and training	Competence, training and awareness	• Training records;
4.3	Employing temporary workers or outsourcing		• Staff record.

TABLE I–15. CORRESPONDENCE BETWEEN THIS PUBLICATION AND LENA QMS (cont.)

		LENA QMS	
Section and title of main text		Process	Documents
2	Management system	Quality Management System	• All documents belonging to LENA's Quality Management System.
3	Management responsibility	Management responsibility	• Management System Manual.
3.1	Management commitment	Management commitment	• Management System Manual.
3.2	Management processes	Quality policy	• Management System Manual;
3.2.2	Planning and reporting	Planning	• LENA quality policy;
3.2.3	Management system administration	Responsibility, authority and communication	• Management System Manual; • Procedure 'management review';
3.2.4	Organizational structure	Management review	• Management review records; • Improvement plan.
4	Resource management	Human resources	• Management System Manual;
4.1	Recruitment, selection and appointment	General	• Procedure 'employee training';
4.2	Performance, assessment and training	Competence, training and awareness	• Training records; • Staff record.
4.3	Employing temporary workers or outsourcing		

FIG. I–1. View of the reactor hall with the reactor block.

FIG. I–2. View of the reactor core.

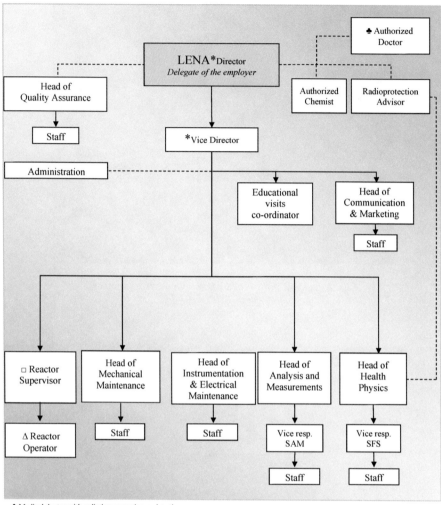

♣ Medical doctor with radiation protection registration
* Personnel in possession of Licence for Technical Management LENA reactor (Technical Director)
□ Personnel in possession of (1nd level) Licence for Reactor Operation
Δ Personnel in possession of (2nd level) Licence for Reactor Operation

FIG. I–3. LENA TRIGA reactor organization chart.

FIG. I–4. Relationship between the various interested parties and the most relevant requirements.

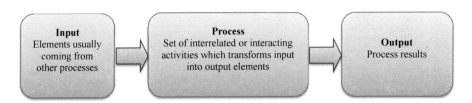

FIG. I–5. Process definition.

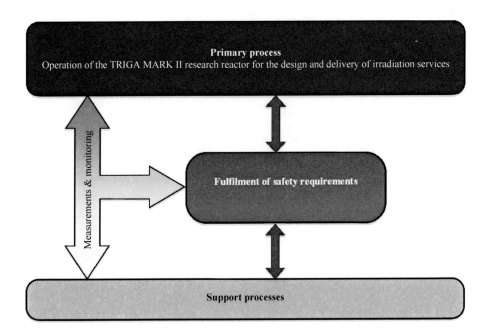

FIG. I–6. Primary and support processes relationship.

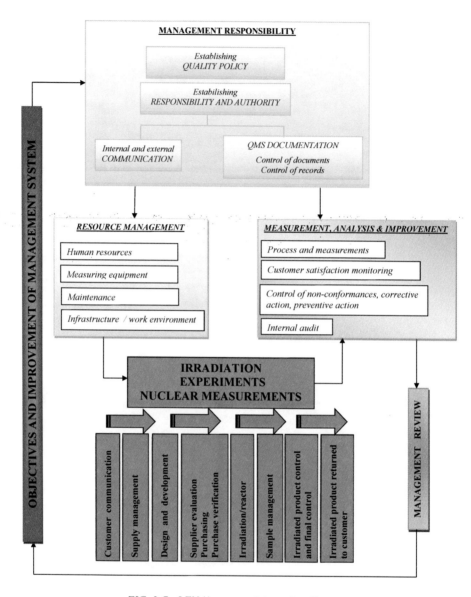

FIG. I–7. LENA's process interaction diagram.

The first step in the LENA QMS building process was gaining senior management's commitment to support the system. LENA management understood the benefits of a LENA QMS and what it would take to put the system in place. Management commitment and vision were clear and communicated across the organization.

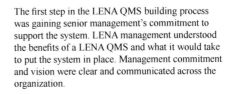

A team with representation from key management functions and expertise areas identified and assessed issues, opportunities and existing processes. The project team included the following functions (see Fig I–3):

- LENA Director;
- Quality assurance;
- Health physics;
- Instrumentation and electrical maintenance;
- Mechanical maintenance;
- Analysis and measures.

As a first step in the implementation process, the incumbents of all the above mentioned functions attended the required training in the LENA QMS. This allowed them to develop their competence in both technical and management issues regarding the QMS implementation process.

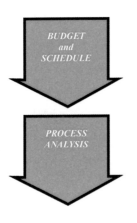

The project team prepared a preliminary budget and schedule for the LENA QMS development. Costs included staff time, training costs, eventual consulting assistance, materials and equipment. The schedule considered, among other things, the various tasks described below.

One of the first tasks the project team completed was an in depth analysis of the organization's processes, evaluating its structure and existing procedures, policies, environmental issues, training programmes and other factors. A gap analysis, along with the determination of the elements of the current system that are still in good shape as well as the ones that need additional work, was performed (see para. I.–2.3).

The LENA Director, in consultation with the Head of Quality Assurance (see organization's chart para. I.–1.2), defined the quality policy and its correlated objectives, ensuring an effective communication and understanding to all employees.

FIG. I–8. Implementation process outline.

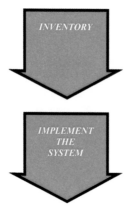

The project team evaluated the existing documentation, procedure, requirements and good practices not already documented. The team defined what had to be implemented, modified or integrated in order to comply with the project purpose and objectives.

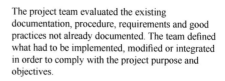

At this point, the team developed procedures and other system documents. In some cases, this involved modifying existing procedures (or adapting others for these specific purposes) and/or developing new procedures. The involvement of employees who had experience with the specific activities was a determining factor for procedures development: Every skill present in the organization contributed to the effectiveness of the process. As soon as a new procedure (or a reviewed document) was approved, it was implemented and the results were recorded. Specific training for all personnel units involved in the specific activity was carried out and its effectiveness recorded.

Both during the LENA QMS implementation process and once it was up and running, assessing system performance was essential. This is continuously carried out through periodic LENA QMS internal audits and ongoing monitoring and measurement. Management review was another important tool for the assessment of LENA QMS performance. Furthermore, this process provides the opportunity for continuous improvement over time.

After its completion, the LENA QMS was certified as conforming to ISO 9001:2008. An independent certification body performed the assessment (external audit) and issued the certificate following approximately two years of work by the project team and all the organization's staff.

FIG. I–8. Implementation process outline (cont.).

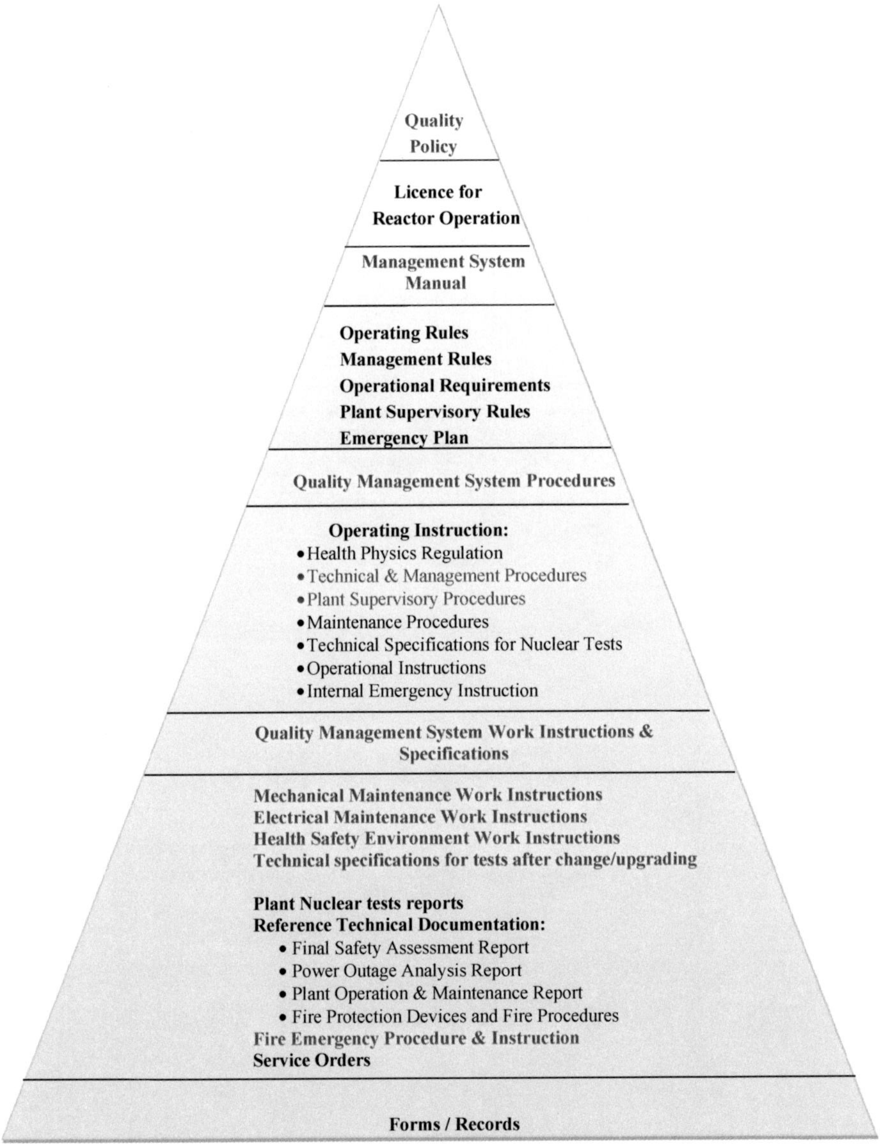

Quality
Policy

**Licence for
Reactor Operation**

Management System
Manual

**Operating Rules
Management Rules
Operational Requirements
Plant Supervisory Rules
Emergency Plan**

Quality Management System Procedures

Operating Instruction:
- Health Physics Regulation
- Technical & Management Procedures
- Plant Supervisory Procedures
- Maintenance Procedures
- Technical Specifications for Nuclear Tests
- Operational Instructions
- Internal Emergency Instruction

**Quality Management System Work Instructions &
Specifications**

**Mechanical Maintenance Work Instructions
Electrical Maintenance Work Instructions
Health Safety Environment Work Instructions
Technical specifications for tests after change/upgrading**

**Plant Nuclear tests reports
Reference Technical Documentation:**
- Final Safety Assessment Report
- Power Outage Analysis Report
- Plant Operation & Maintenance Report
- Fire Protection Devices and Fire Procedures

**Fire Emergency Procedure & Instruction
Service Orders**

Forms / Records

*FIG. I–9. LENA document system after management system implementation.
Newly issued documents are highlighted in green.*

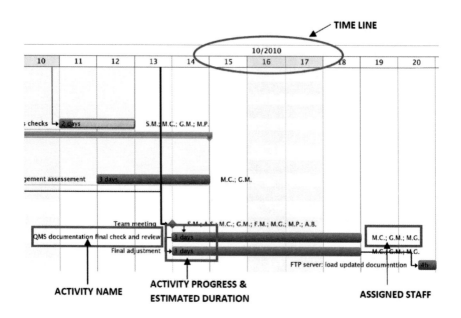

FIG. I–10. Screenshot of LENA Gantt chart.

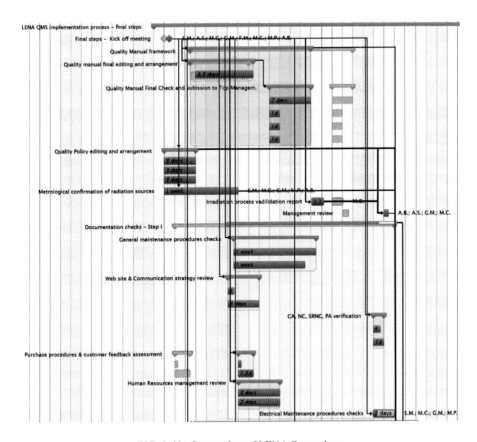

FIG. I–11. Screenshot of LENA Gantt chart.

137

#	Info	Title	Given Planned Work	Giv. Plan. Dur.	# Predecessors	% Complete	Assigned Resource Initials
0	⊕⊖	▼ 📁 LENA QMS implementation process – final steps				99%	C.P.; M.P.; M.C.; ...
1	⊕⊙⊖	▷ Final steps – Kick off meeting				100%	S.M.; A.S.; M.C.; ...
10	⊘	▼ Quality Manual framework			1	100%	M.C.; A.B.; M.G.; ...
11	⊕⊖	▷ Quality manual final editing and arrangement		8.5 days	1	100%	M.C.
13	⊘	▷ Quality Manual Final Check and subission to Top Managem.		2 days	11	100%	M.C.; G.M.; M.G....
18	⊕⊖	▷ Quality Policy editing and arrangement		3 days	1	100%	A.S.; M.C.; A.B.
22	⊕⊖	▷ Metrological confirmation of radiation sources		5 days	1	100%	S.M.; M.C.; G.M. ...
28	⊕⊙⊖	▷ Irradiation process vadilidation report	6 days		1	100%	M.C.
30	⊘	▷ Management review	2 days	100 %	18; 28	100%	A.B.; A.S.; G.M.; ...
35	⊕⊖	▼ Documentation checks – Step I				95%	C.P.; M.P.; M.C.; ...
36	⊕⊖	▷ General maintenance procedures checks		5 days	1	100%	M.C.; F.M.
39	⊕⊖	▷ Web site & Communication strategy review		3 days	1	100%	M.C.; G.V.
42	⊕⊖	▷ CA, NC, SRNC, PA verification	3 days		1	100%	G.M.; M.C.
45	⊘	▷ Purchase procedures & customer feedback assessment		3 days	1	100%	M.C.; C.P.
48	⊕⊖	▷ Human Resources management review	0.5 days	1 day	1	100%	M.C.; G.M.
51	⊘	▷ Electrical Maintenance procedures checks	100 %	2 days	1	50%	S.M.; M.C.; G.M. ...
56	⊘	▼ Documentation cheks – Step II	10 days	10 days	1	100%	M.C.; M.P.; A.B.; ...
57	⊕⊙⊖	▷ Work Instructions final editing and submission	5 days			100%	G.M.; M.P.
60	⊘	▷ Work Instructions final approval	1 day		57	100%	M.C.; A.S.; A.B.
64	⊕⊖	▷ Records management assessement	3 days	3 days		100%	M.C.; G.M.
67	⊘	▼ Complete still– pending activities reported from audit			1	100%	M.C.; G.M.
68	⊘	▷ Customenr's satisfaction evaluation procedure final check		5 days	1	100%	M.C.; G.M.

FIG. I–12. Screenshot of LENA activity schedule.

1. **Scope**: Cooling system pumps preventive maintenance

2. **Application field:** Reactor cooling system, centrifugal pumps:

> • Primary cooling circuit — heat exchanger #1
> • Secondary cooling circuit — heat exchanger #1 and #2
> • Refilling feeding pumps (pool and tank)
> • Thermalizing pool circulating pump

3. **Responsibility:** Mechanical Maintenance Section (SMM)

4. **Work instruction:**

4.1 Equipment position

All equipment is labelled for a clear and easy identification:

Pump function	Collocation	Label	Label colour
Primary cooling circuit	Heat exchanger	PP1	Red
		PP2	
Secondary cooling circuit	Heat exchanger room	PS1	Orange
		PS2	
Refilling feeding pumps	Heat exchanger room	PC1	Blue
Pool circulating pump	Thermalizing pool	PCP 1	Gray

4.2 Frequency: Every 15 days

4.3 Operating instructions

Notify reactor supervisor about the start of the maintenance works; the reactor must be in shutdown condition.

Be sure to disconnect power supply before maintenance works. Wear the necessary personal protective equipment as indicated in the Health physics regulation.

NOTE: internal parts of primary cooling circuit pumps PP1 and PP2 are subjected to radioactive contamination. Please consult Health physics safety instruction before start of the work.

(…..........................…)

For each pump, the following test and verification must be carried out:

> • Lubricating level check
> • Bearings greasing (if necessary)
> • Water leakage check
> • Nuts tightening
> • Check for corrosion of metal parts
> • Selection of the pumps in operation/standby pumps (only twin pumps) to be done on the switchboard

(…..........................…)

5. **Records:** The results of the operation shall be recorded on the SMMCM form.

6. **References**

– Mechanical maintenance and operating manual, Ch. 4: water system
– Pumps maintenance manual
– Health physics regulations
(…..........................…)

FIG. I–13. A LENA work instruction.

139

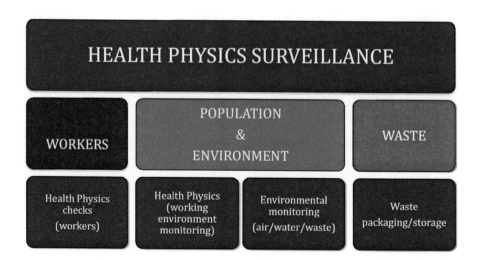

FIG. I–14. Overview of health physics surveillance programme.

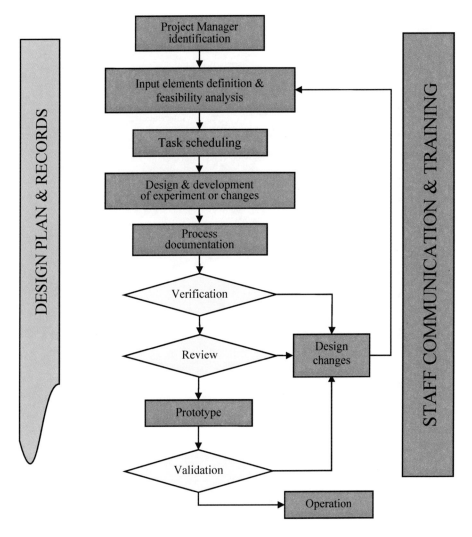

FIG. I–15. Flow chart used to plan specific activities (such as changes in an experiment already in progress or the planning of a new experiment) not covered by the standard procedures.

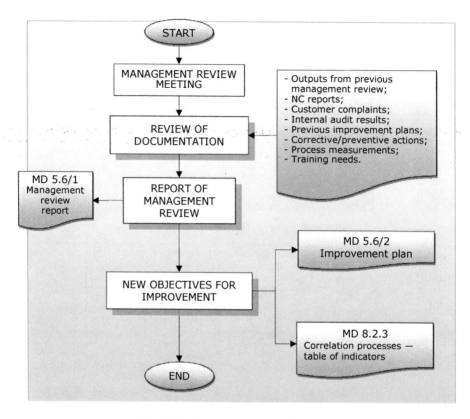

FIG. I–16. Management review flow chart.

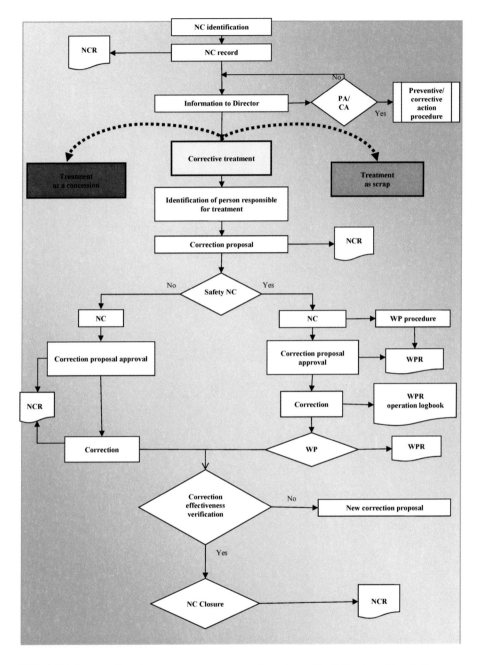

FIG. I–17. Non-conformity management process flow chart. The following abbreviations are used in Fig. I–17: NCR — Non-conformance report; WP — Work Permission; WPR — Work Permission Report.

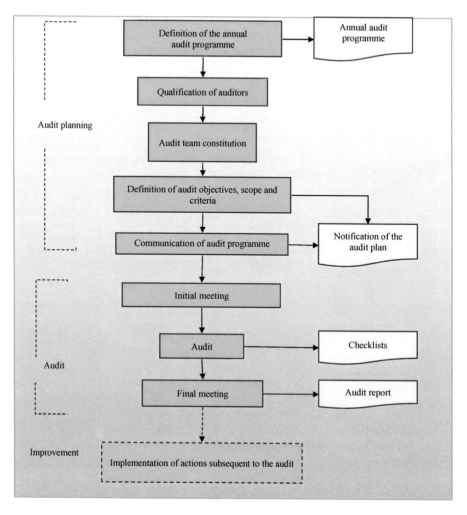

FIG. I–18. Internal audit management process flow chart.

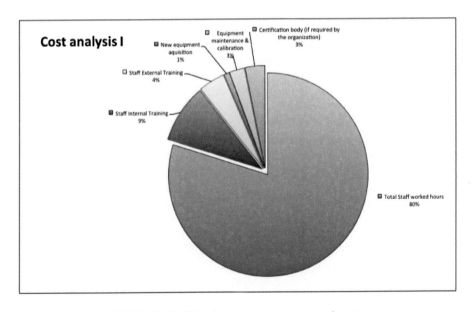

FIG. I–19. Staff involvement costs versus total costs.

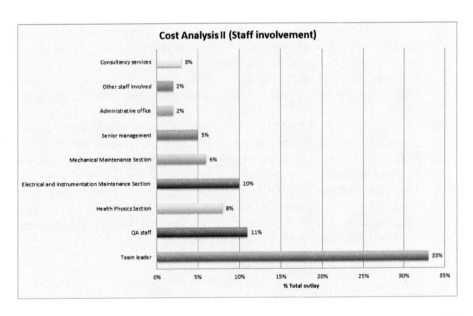

FIG. I–20. Staff involvement (in percentage and for function) in the LENA QMS implementation process.

REFERENCES TO ANNEX I

[I–1]   Code of Conduct on the Safety of Research Reactors, Resolution GOV/OR.1088, IAEA, Vienna (2004).

[I–2]   Treaty on the Non-proliferation of Nuclear Weapons, INFCIRC/140, IAEA, Vienna (1970).

[I–3]   Convention on the Physical Protection of Nuclear Material, INFCIRC/274/Rev.1, IAEA, Vienna (1980).

[I–4]   INTERNATIONAL ATOMIC ENERGY AGENCY, Safety of Research Reactors, IAEA Safety Standards Series No. NS-R-4, IAEA, Vienna (2005).

[I–5]   INTERNATIONAL ATOMIC ENERGY AGENCY, The Management System for Facilities and Activities, IAEA Safety Standards Series No. GS-R-3, IAEA, Vienna (2006).

[I–6]   INTERNATIONAL ATOMIC ENERGY AGENCY, Application of the Management System for Facilities and Activities, IAEA Safety Standards Series No. GS-G-3.1, IAEA, Vienna (2006).

[I–7]   INTERNATIONAL ATOMIC ENERGY AGENCY, The Management System for Nuclear Installations, IAEA Safety Standards Series No. GS-G-3.5, IAEA, Vienna (2009).

[I–8]   INTERNATIONAL ATOMIC ENERGY AGENCY, Maintenance, Periodic Testing and Inspection of Research Reactors, IAEA Safety Standards Series No. NS-G-4.2, IAEA, Vienna (2006).

[I–9]   INTERNATIONAL ATOMIC ENERGY AGENCY, Optimization of Research Reactor Availability and Reliability: Recommended Practices, IAEA Nuclear Energy Series No. NP-T-5.4, IAEA, Vienna (2008).

[I–10] INTERNATIONAL ORGANIZATION FOR STANDARDIZATION, Quality Management Systems — Fundamentals and Vocabulary, ISO 9000:2005, ISO Geneva (2005).

[I–11] INTERNATIONAL ORGANIZATION FOR STANDARDIZATION, Quality Management Systems — Requirements, ISO 9001:2008, ISO Geneva (2008).

[I–12] INTERNATIONAL ORGANIZATION FOR STANDARDIZATION, General Requirements for the Competence of Testing and Calibration Laboratories, ISO/IEC 17025:2005, ISO, Geneva (2005).

[I–13] INTERNATIONAL ORGANIZATION FOR STANDARDIZATION, Measurement Management Systems — Requirements for Measurement Processes and Measuring Equipment, ISO 10012:2003. ISO, Geneva (2003).

EXAMPLE OF A GRADED APPROACH TO THE APPLICATION OF MANAGEMENT SYSTEM REQUIREMENTS FOR A RESEARCH REACTOR

II–1. INTRODUCTION

The research reactor installation is divided into structure, systems and components, each comprising items, services and processes. Factors with a significance to nuclear safety, reliability, complexity, design and experience are determined and a quality grade (A, B, C, D) is assigned to each structure, system or component. The quality grade is obtained by applying the results of a qualification formula and the criteria from Assignation of Quality Grades described below. The management system defines applicable requirements for each quality grade.

II–2. QUALIFICATION FORMULA

A 'total quality rating' of the structure, systems and components is obtained by assigning the values for each of the factors considered in the formula. The criteria applied to obtain the different values for each factor are not discussed here.

Total quality rating $(TQR) = 2a + b + c + d + e$

The TQR may refer to a general system or to its components because the components of a system will not necessarily have the same quality class as the system itself.

A brief description of the factors:

Safety (a): This factor includes nuclear, radiation, physical and what is known as industrial safety. It has a weight of 2 and its value can range from 0 to 5, with 5 having the highest safety relevance.

Reliability (b): This factor includes consideration of possible loss of profit, e.g. through delay, failed repair work, interruption of service i.e. radioisotope production or the irradiation of an experiment, etc. Its values can range from 0 to 5, with 5 representing the greatest threat to reliability.

Complexity (c): This factor includes consideration of the complexity of the design, difficulties in replacing parts, accessibility for maintenance and unique structures, systems and components design. Its values can range from 0 to 5, with 5 representing the greatest complexity.

Design State (d): This factor gives consideration to identifying the maturity of the design, from fully tested structures, systems and components design of proven quality able to be used without modifications, to a new design to be developed from basic principles and data. Whenever a prototype is to be built, this action will be valued by assigning a lower factor to it. Its values can range from 0 to 5, with 5 representing the lowest degree of maturity.

Experience (e): This factor takes into account the accumulated and objective experience of the structures, systems and components, obtained by the organization, by suppliers, by other organizations or by recognized consultants and/or contractors. Its values can range from 0 to 5, with 5 representing the lowest level of experience.

II–3. ASSIGNATION OF QUALITY GRADES

Four quality grades are used: A, B, C and D. Quality grade A represents the most stringent level of requirements (Table II–1).

II–4. EXAMPLES

Tables II–2–II–4 give some examples of the requirements of each of the four quality grades.

TABLE II–1. ASSIGNATION OF QUALITY GRADES

Quality grades	Assignment criteria	
A	Items with factor:	a = 4 or 5
	Items with factor:	b = 5
	Items with:	TQR = 25–30
B	Items with factor:	a = 2 or 3
	Items with factor:	b = 3 or 4
	Items with:	TQR = 18–24
C	Items with:	TQR = 5–17
D	Items with:	TQR = 0–4

TABLE II–2. QUALIFICATION AND TRAINING QUALITY GRADE REQUIREMENTS

Qualification and training	Quality grade			
	A	B	C	D
Qualification				
Regulatory authority licence for reactor commissioning, reactor operation, radiation safety; manufacturing of nuclear fuel element	•	•		
Certification for design and manufacturing	•	•		
Welding and welder qualification/welding inspectors	•	•		
Non-destructive tests	•	•	•	
Training				
All personnel involved in design control, manufacturing, installation, startup, commissioning; operation, maintenance, test and inspections	•	•		
Internal auditors	•	•		
Quality officers	•	•	•	
Personnel	•	•	•	

TABLE II–3. PROCUREMENT QUALITY GRADE REQUIREMENTS

Procurement	Quality grade			
	A	B	C	D
Supplier evaluation and selection (prior to the awarding of the procurement order or contract)	•	•		
Inspection of the supplier's facility by the technical representative or quality officer	•	•	•	
Document evidence from the supplier that the procured items meet procurement quality requirements, such as codes, standards, or specifications	•	•	•	•
Periodic verification of the supplier's certificates of conformance to assure their continuing validity	•	•		
Evaluation of the performance of the supplier with the participation of the technical representative/procurement department/quality division	•	•		

TABLE II–4. NON-CONFORMANCES AND CORRECTIVE ACTIONS QUALITY GRADE REQUIREMENTS

Non-conformances and corrective actions	Quality grades			
	A	B	C	D
Non-conformances				
Components that do not conform to requirements reviewed and approved by the designer and the corresponding management grades	•	•	•	•
Indication of the decision taken: 'use as is' / 'repair' / 're-work'	•	•	•	
Analyses of reports	•	•		
Corrective actions				
Promptly identified and corrected (failures, malfunctions, deficiencies, deviations, defective material and equipment)	•	•		
In the case of conditions significantly adverse to quality, the cause is determined and a corrective action is taken to preclude repetition. This is further documented and reported to the corresponding management grades	•	•		

Annex III

CONFIGURATION MANAGEMENT

This example describes a process methodology for establishing consistency between design requirements, physical configuration and documentation for the facility and maintaining this consistency throughout its life cycle (design, bidding, procurement, construction, commissioning, maintenance, operation and decommissioning).

(1) Senior management (or division manager):
    — Communicates the need for and the concept of configuration management in order to ensure that a culture of configuration management exists at all organizational levels.

(2) Line manager (projects, facilities and departments):
    — Prepares a configuration management programme (or plan) in which design activities are initiated and properly extended to other phases as work progresses.
    — Includes in the abovementioned programme requirements for external contractors and suppliers to ensure that configuration management principles are adopted in checking the results of their activities.
    — Integrates all the key activities as presented in the list Key Activities of a Configuration Management Programme, below, or in other relevant processes and procedures.
    — Clearly identifies, via documented procedures, the design authority during all phases of the life cycle of the facility. Includes the responsibilities for maintaining design related information and knowledge of how to use this information. The main responsibility of the design authority is the approval of major design changes to the baseline design configuration that underpins the safety analysis report. This could be done as chair of a committee looking into the changes.
    — Ensures that the procurement, construction, operation, maintenance, decommissioning and testing of the physical facility are in accordance with the design requirements outlined in the design documentation and that this consistency is maintained throughout the life of the facility. A procedure to report, review and approve or reject non-conformances has to be in place before construction starts.
    — Ensures that the facility configuration baseline, which includes as-built drawings and operating (including maintenance) procedures, accurately reflects both the physical configuration and the design requirements under configuration management.

— Ensures that the project/facility procedures guarantee that, after completion of maintenance and modification activities, all equipment is strictly returned to the approved operating configuration and the related documentation is updated to reflect field conditions.

— Implements an integrated configuration management system, which includes as a minimum document record and equipment information management functions to achieve the objectives of the configuration management programme. The use of computerized software applications will ideally be considered whenever feasible.

Key Activities of a Configuration Management Programme

(a) Configuration identification
  • Selection of physical configuration;
  • Identification of facility configuration information;
  • Formatting of documentation;
  • Storage of documentation.
(b) Change control
  • Minimize design changes;
  • Control design changes;
  • Document as-built configuration;
  • Transfer of design authority.
(c) Record and maintain configuration status
  • Identification conventions;
  • Updating of design documentation;
  • Operational configuration;
  • Maintenance of design documentation for facility.
(d) Configuration audits and independent verification
  • Configuration audits;
  • Independent verification.

Annex IV

EXAMPLES OF A SYSTEM HEALTH PROCESS

This example describes a process methodology enabling facilities to increase safety and reliability and to meet regulatory requirements on performance monitoring. An implemented system health process will ensure that:

— General plant conditions, equipment deficiencies and performance data are known;
— The results of testing and trending are evaluated;
— Changes and improvements are recommended;
— Improvement initiatives and corrective actions improve the overall safety and reliability of the facility.

A system health process can be generalized into the steps shown in Table IV–1. The flow diagram of Fig. IV–1 provides a visual representation of the system health process.

TABLE IV–1. STEPS OF A SYSTEM HEALTH PROCESS

Person responsible	Step	Tasks and outputs
Process owner (or manager)	(1)	Set up interfaces and channels for decision making and oversight; Identify system responsible expert (SRE), system health team (SHT) and system health oversight committee (SHOC).
Expert responsible for the system	(2)	Prepare and maintain a system archive.
	(3)	Prepare a system performance monitoring plan (SPMP) that includes: • System description and list of the major components/ equipment; • Goals for system performance; • A functional failure analysis and failure mode analysis to document component failures that could render the system incapable of carrying out its design functions; • Detailed ageing management plan; • Walk-down checklist; • Obsolescence strategies; • Performance indicators for the system.
	(4)	Ensure monitoring activities are performed.
Expert responsible for the system with input from system health team	(5)	Prepare system performance monitoring report (SPMR) as scheduled that summarizes the system performance results and activities during the reporting period, as well as details of significant trends and system performance issues.
Process owner and system health oversight committee	(6)	Assess the effectiveness of the system health process and recommend changes for improvement.

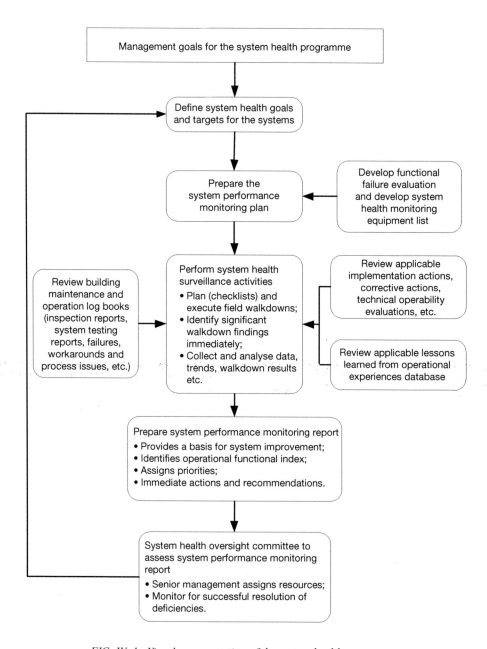

FIG. IV–1. Visual representation of the system health programme.

Annex V

EXAMPLE OF AN AGEING MANAGEMENT PROCESS

This example process describes a methodology for an ageing management programme[1] which is used to ensure that structures, systems and components in nuclear facilities comply with required safety functions throughout the entire life of the facility, within acceptable limits defined by the design basis and by safety analysis, with account taken of changes that occur over time and with use. The process includes engineering, operational, inspection and maintenance activities, and is designed to improve system performance by decreasing ageing related malfunctions and critical failures.

The process consists of the following five steps:

(1) *Screening for critical systems and components.*
     This step is to identify and prioritize the critical structures, systems and components within the facility by using a risk graded approach. It may utilize categories whereby structures, systems and components considered critical are categorized as '1'; those considered non-critical fall into category '2' and those that are designated run-to-failure are category '3'. The criteria for determining a category for the structures, systems and components are based on their safety function, the impact of failure on facility, public, worker and environmental safety, and production, as specified in the screening procedure.

(2) *Ageing assessments on critical aspects of the facility.*
     This step evaluates whether ageing degradation effects, as revealed by the operational and maintenance history, have compromised the design basis or the functional, safety and operational requirements of the structures, systems or components. It can be conducted either by condition assessments or by life assessments. The methodology entails a general review of plant data to determine the current condition of components and to evaluate ageing degradation at the component level. It provides a prognosis for attainment of design life and/or long term operation. The recommendations made in the condition or life assessments provide the technical basis for on-going ageing management of the structures, systems and components.

---

[1] Detailed guidance on ageing management for research reactors can be found in: INTERNATIONAL ATOMIC ENERGY AGENCY, Use of a Graded Approach in the Application of the Safety Requirements for Research Reactors, IAEA Safety Standards Series No. SSG-22, IAEA, Vienna (2012).

(3) *Development of a structures, systems and component specific ageing management plan.*

Ageing is a process in which characteristics of structures, systems and components gradually change over time or with use. This process may proceed by way of a single ageing mechanism or a combination of several ageing mechanisms. As a result, it could impare the ability of a structure, system or component to function within its acceptance criteria.

An ageing management plan is drawn up to provide strategies to effectively manage the ageing related degradation mechanisms of structures, systems and components identified in associated condition or life assessment reports and in other resources that are part of the ageing assessment. Other resources are employed to mitigate the consequences of ageing as far as this is possible.

The plan contains two main parts:

— Recommendations to manage ageing and mitigate failures;
— A detailed plan chart with methods to manage the ageing related degradation mechanisms of structures, systems and components identified in the condition or life assessment.

(4) *Implementation.*

Implementation starts with recommendations made in the plan. Specifically, recommendations are reviewed, classified and approved or rejected. Those approved are implemented and monitored with appropriate records kept.

(5) *Continuous improvement.*

This step includes the review of operating performance, inspection and maintenance histories, event reports, information from results of research and development, self-assessment and operating experience.

The structure, system and component specific ageing management plans are updated based on the data collected. Revisions could be necessary in any of the following situations:

— New ageing related degradation mechanisms are discovered;
— Deficiencies in current procedures used in the ageing management activities;
— Structures, systems and components are modified or replaced;
— The production of a new condition or life assessment;
— New information or improved tools and methods have been discovered that would have a positive impact on the managing of ageing within the system.

Responsibilities (The first two cover the link to the system health process):

System health oversight committee. Committee responsible for reviewing and approving recommendations to mitigate or manage ageing.

System health team. Responsible for reviewing, classifying and prioritizing recommendations, and incorporating these recommendations into system performance monitoring reports.

Ageing assessment teams. Responsible for conducting ageing assessments on structures, systems and components that will be included in the plan and for making recommendations to effectively manage the ageing of these structures, systems and components.

Integrated expert panel. Responsible for performing screening to identify and prioritize critical structures, systems and components to determine the scope to be covered by the ageing management plan.

Experts responsible for the system. Responsible for developing and maintaining the ageing management plan, giving recommendations on emerging ageing issues within their systems, ensuring that approved recommendations are being carried out and verifying completion of these recommendations.

Facility management. Responsible for supporting screening processes, implementation (ensuring that necessary surveillance, monitoring, maintenance and inspection activities are being carried out), and taking actions at an early stage on any ageing issues that put facility safety and/or operation at risk.

CONTRIBUTORS TO DRAFTING AND REVIEW

Bernido, C.	PNRI, Philippines
Boado Magan, J.	International Atomic Energy Agency
Boogaard, J.P.	International Atomic Energy Agency
Bradley, E.E.	International Atomic Energy Agency
Cagnazzo. M.	University of Pavia, Italy
Cirimello, R.	INVAP, Argentina
Dahlgren, K.	Vattenfall, Sweden
Kekki, T.	VTT, Finland
Lavoier, R.	AECL Nuclear Laboratories, Canada
Magrotti G.	University of Pavia, Italy
Mallants, F.	SCK•CEN, Belgium
Pieroni, N.	International Atomic Energy Agency
Redman, N.	Amethyst Management Limited, United Kingdom
Ridikas, D.	International Atomic Energy Agency
Shokr, A.	International Atomic Energy Agency
Vincze, P.	International Atomic Energy Agency
Wijtsma, F.	Nuclear Research and Consultancy Group, Netherlands

IAEA
International Atomic Energy Agency

Where to order IAEA publications

In the following countries IAEA publications may be purchased from the sources listed below, or from major local booksellers. Payment may be made in local currency or with UNESCO coupons.

AUSTRALIA
DA Information Services, 648 Whitehorse Road, MITCHAM 3132
Telephone: +61 3 9210 7777 • Fax: +61 3 9210 7788
Email: service@dadirect.com.au • Web site: http://www.dadirect.com.au

BELGIUM
Jean de Lannoy, avenue du Roi 202, B-1190 Brussels
Telephone: +32 2 538 43 08 • Fax: +32 2 538 08 41
Email: jean.de.lannoy@infoboard.be • Web site: http://www.jean-de-lannoy.be

CANADA
Bernan Associates, 4501 Forbes Blvd, Suite 200, Lanham, MD 20706-4346, USA
Telephone: 1-800-865-3457 • Fax: 1-800-865-3450
Email: customercare@bernan.com • Web site: http://www.bernan.com

Renouf Publishing Company Ltd., 1-5369 Canotek Rd., Ottawa, Ontario, K1J 9J3
Telephone: +613 745 2665 • Fax: +613 745 7660
Email: order.dept@renoufbooks.com • Web site: http://www.renoufbooks.com

CHINA
IAEA Publications in Chinese: China Nuclear Energy Industry Corporation, Translation Section, P.O. Box 2103, Beijing

CZECH REPUBLIC
Suweco CZ, S.R.O., Klecakova 347, 180 21 Praha 9
Telephone: +420 26603 5364 • Fax: +420 28482 1646
Email: nakup@suweco.cz • Web site: http://www.suweco.cz

FINLAND
Akateeminen Kirjakauppa, PO BOX 128 (Keskuskatu 1), FIN-00101 Helsinki
Telephone: +358 9 121 41 • Fax: +358 9 121 4450
Email: akatilaus@akateeminen.com • Web site: http://www.akateeminen.com

FRANCE
Form-Edit, 5, rue Janssen, P.O. Box 25, F-75921 Paris Cedex 19
Telephone: +33 1 42 01 49 49 • Fax: +33 1 42 01 90 90
Email: formedit@formedit.fr • Web site: http://www. formedit.fr

Lavoisier SAS, 145 rue de Provigny, 94236 Cachan Cedex
Telephone: + 33 1 47 40 67 02 • Fax +33 1 47 40 67 02
Email: romuald.verrier@lavoisier.fr • Web site: http://www.lavoisier.fr

GERMANY
UNO-Verlag, Vertriebs- und Verlags GmbH, Am Hofgarten 10, D-53113 Bonn
Telephone: + 49 228 94 90 20 • Fax: +49 228 94 90 20 or +49 228 94 90 222
Email: bestellung@uno-verlag.de • Web site: http://www.uno-verlag.de

HUNGARY
Librotrade Ltd., Book Import, P.O. Box 126, H-1656 Budapest
Telephone: +36 1 257 7777 • Fax: +36 1 257 7472 • Email: books@librotrade.hu

INDIA
Allied Publishers Group, 1st Floor, Dubash House, 15, J. N. Heredia Marg, Ballard Estate, Mumbai 400 001,
Telephone: +91 22 22617926/27 • Fax: +91 22 22617928
Email: alliedpl@vsnl.com • Web site: http://www.alliedpublishers.com

Bookwell, 2/72, Nirankari Colony, Delhi 110009
Telephone: +91 11 23268786, +91 11 23257264 • Fax: +91 11 23281315
Email: bookwell@vsnl.net

ITALY
Libreria Scientifica Dott. Lucio di Biasio "AEIOU", Via Coronelli 6, I-20146 Milan
Telephone: +39 02 48 95 45 52 or 48 95 45 62 • Fax: +39 02 48 95 45 48
Email: info@libreriaaeiou.eu • Website: www.libreriaaeiou.eu

JAPAN
Maruzen Company Ltd, 1-9-18, Kaigan, Minato-ku, Tokyo, 105-0022
Telephone: +81 3 6367 6079 • Fax: +81 3 6367 6207
Email: journal@maruzen.co.jp • Web site: http://www.maruzen.co.jp

REPUBLIC OF KOREA
KINS Inc., Information Business Dept. Samho Bldg. 2nd Floor, 275-1 Yang Jae-dong SeoCho-G, Seoul 137-130
Telephone: +02 589 1740 • Fax: +02 589 1746 • Web site: http://www.kins.re.kr

NETHERLANDS
De Lindeboom Internationale Publicaties B.V., M.A. de Ruyterstraat 20A, NL-7482 BZ Haaksbergen
Telephone: +31 (0) 53 5740004 • Fax: +31 (0) 53 5729296
Email: books@delindeboom.com • Web site: http://www.delindeboom.com

Martinus Nijhoff International, Koraalrood 50, P.O. Box 1853, 2700 CZ Zoetermeer
Telephone: +31 793 684 400 • Fax: +31 793 615 698
Email: info@nijhoff.nl • Web site: http://www.nijhoff.nl

Swets and Zeitlinger b.v., P.O. Box 830, 2160 SZ Lisse
Telephone: +31 252 435 111 • Fax: +31 252 415 888
Email: infoho@swets.nl • Web site: http://www.swets.nl

NEW ZEALAND
DA Information Services, 648 Whitehorse Road, MITCHAM 3132, Australia
Telephone: +61 3 9210 7777 • Fax: +61 3 9210 7788
Email: service@dadirect.com.au • Web site: http://www.dadirect.com.au

SLOVENIA
Cankarjeva Zalozba d.d., Kopitarjeva 2, SI-1512 Ljubljana
Telephone: +386 1 432 31 44 • Fax: +386 1 230 14 35
Email: import.books@cankarjeva-z.si • Web site: http://www.cankarjeva-z.si/uvoz

SPAIN
Díaz de Santos, S.A., c/ Juan Bravo, 3A, E-28006 Madrid
Telephone: +34 91 781 94 80 • Fax: +34 91 575 55 63
Email: compras@diazdesantos.es, carmela@diazdesantos.es, barcelona@diazdesantos.es, julio@diazdesantos.es
Web site: http://www.diazdesantos.es

UNITED KINGDOM
The Stationery Office Ltd, International Sales Agency, PO Box 29, Norwich, NR3 1 GN
Telephone (orders): +44 870 600 5552 • (enquiries): +44 207 873 8372 • Fax: +44 207 873 8203
Email (orders): book.orders@tso.co.uk • (enquiries): book.enquiries@tso.co.uk • Web site: http://www.tso.co.uk

On-line orders
DELTA Int. Book Wholesalers Ltd., 39 Alexandra Road, Addlestone, Surrey, KT15 2PQ
Email: info@profbooks.com • Web site: http://www.profbooks.com

Books on the Environment
Earthprint Ltd., P.O. Box 119, Stevenage SG1 4TP
Telephone: +44 1438748111 • Fax: +44 1438748844
Email: orders@earthprint.com • Web site: http://www.earthprint.com

UNITED NATIONS
Dept. I004, Room DC2-0853, First Avenue at 46th Street, New York, N.Y. 10017, USA
(UN) Telephone: +800 253-9646 or +212 963-8302 • Fax: +212 963-3489
Email: publications@un.org • Web site: http://www.un.org

UNITED STATES OF AMERICA
Bernan Associates, 4501 Forbes Blvd., Suite 200, Lanham, MD 20706-4346
Telephone: 1-800-865-3457 • Fax: 1-800-865-3450
Email: customercare@bernan.com · Web site: http://www.bernan.com

Renouf Publishing Company Ltd., 812 Proctor Ave., Ogdensburg, NY, 13669
Telephone: +888 551 7470 (toll-free) • Fax: +888 568 8546 (toll-free)
Email: order.dept@renoufbooks.com • Web site: http://www.renoufbooks.com

Orders and requests for information may also be addressed directly to:

Marketing and Sales Unit, International Atomic Energy Agency
Vienna International Centre, PO Box 100, 1400 Vienna, Austria
Telephone: +43 1 2600 22529 (or 22530) • Fax: +43 1 2600 29302
Email: sales.publications@iaea.org • Web site: http://www.iaea.org/books